ECO AGRI REVOLUTION

About the Author

Dr. M.H. Mehta (Munibhai) is President of The Science Ashram and Chairman of Gujarat Life Sciences. He was Vice Chancellor of Gujarat Agricultural University and before that Executive Director - GSFC Ltd., Senior Scientist - Bhabha Atomic Research Centre and visiting Scientist-University of California and Arizona. He was founder Director of GSFC Science Foundation and GGRC (Gujarat Green Revolution Co.). He holds B.Tech and Ph.D in Chemical Engineering from I.I.T. Bombay.

Dr. Mehta has been awarded Padmashri by Govt. of India. He has been honoured by Indian Science Congress. He was given Agriculture Leadership Award (2012), Life Time Achievement Awards by Asia Pacific Desalination & Water Reuse Association and Indian Membrane Society and earlier the most outstanding Young Chemical Engineer Award by I.I.Ch.E. in 1986. He has been honoured/ invited at several International Institutes/Universities such as Cornell University and University of Arizona (USA), Royal Institute (Stockholm) and recently at Beijing University.

His 20:20 model for Eco Agriculture is increasingly regarded as the practical and apt model for the coming Eco Agri Revolution and water management. He also leads Vaho Vishwamitri Abhiyan - a model holistic plan for the revival of river.

Dr. Mehta has published more than 200 papers, holds 16 patents and has written 4 books. His last book - Science, Spirituality and the World Tomorrow was released by Dr. APJ Abdul Kalam in 2013. A keen sportsman, his other interests are poetry and music and he plays the instrument Kastha Tarang.

ECO AGRI REVOLUTION
Practical Lessons and the Way Ahead

M.H. Mehta

NEW INDIA PUBLISHING AGENCY
New Delhi – 110 034

NEW INDIA PUBLISHING AGENCY
101, Vikas Surya Plaza, CU Block, LSC Market
Pitam Pura, New Delhi 110 034, India
Phone: + 91 (11) 27 34 17 17 Fax: + 91 (11) 27 34 16 16
Email: info@nipabooks.com
Web: www.nipabooks.com

Feedback at feedbacks@nipabooks.com

ISBN: 978-93-86546-19-7

Composed, Designed and Printed in India

To

My wife Ila
my Children, Grand daughters and all
Future generations

M.S. Swaminathan Research Foundation

M.S. Swaminathan
Founder Chairman
Ex-Member of Parliament (Rajya Sabha)

Foreword

We have entered the Anthropocene era where human action will play an important role in shaping the course of development and ecological well being. We need to promote eco technologies which integrate the principles of ecology in technology development and dissemination. We also need to promote production and use of biological software for sustainable agriculture and food security. Dr. Mehta has shown the way on how we can achieve advances in productivity in perpetuity without ecological harm, Therefore, this book is extremely important and very timely.

For every human being, today is a reality and tomorrow is a possibility. The hungry need food today and not the promises for tomorrow. Hence in my earlier book' "From Rio de Janeiro to Johannesburg - Action Today and Not Just Promises for Tomorrow", I had emphasized the great need for "Do How" and "Demonstrated Actions". It is, therefore, heartening to see this book full of 'actions' and demonstrations. The 'Do Ecology' is shown with much conviction. "Do Ecology" is meeting the needs of the current and future generations without ecological harm. 'Do Ecology' involves scientific methods of checkmating potential damage to our environmental assets without any adverse impact on the lives and livelihoods of tribal and rural families and small and marginal farmers.

The author has combined compassion and deep rooted concern for environment and downtrodden people to take up a holistic approach for technology, human and biosphere. The test of our progress is not whether we add more to the abundance of those who have much; it is whether we provide enough for those who have little. Translated into action, this has meant concentration on improving the productivity of small farm holdings and socially under-privileged women and men who toil day and night in sun and rain to produce food for their fellow human beings. Hunger and deprivation can certainly be eliminated sooner than most people consider feasible, provided there is a synergy among technology, public policy and social action. This is the major lesson we can learn from the farm revolution that transferred India's agricultural destiny during the last part of the century. This will also be the base for the coming Eco Agri Revolution as discussed and demonstrated in this book.

The author's pioneering work and leadership in the area of development and commercialization of microbial fertilizers, Bio Pesticides, Bio Composts and farm waste conversion in the past few decades are brought out with illustrated examples. The new generation agri bio inputs are globally emerging as key alternative inputs which are cost - effective, eco-friendly and the most important need of the farmers. Dr. Mehta has been a global leader in this area and through his - The Science Ashram and Gujarat Life Sciences has given a practical and concrete model. What is more valuable is the way this practical and balanced model been taken to farms in different parts of India and several countries in Africa and Asia. The farm level demonstrations are the most important and effective way of convincing the farmers and for technology transfer what Norman Borlough used to emphasize as "Take it to the Farmer!" approach. The 'Antyodaya' concept propounded by Mahatma Gandhi involves a bottom line approach, starting with the poorest person. This seems to be the approach here in organizing demonstrations in the fields of farmers with small holdings in different agro climatic conditions.

Following the severe earth quake in Kutch - Gujarat, where also evapo - transpiration exceeds precipitation in most months of the year, Dr. Mehta's team of Gujarat Agricultural University launched a low cost green house program which greatly helped to open new vistas in water - saving and rehabilitation. This model was recommended in other similar situations in Nepal, Indonesia etc. Similarly, the Bio shield programs discussed for Kalpakam in Tamul Nadu, Uttarkashi in Himalaya and as a part of River Revival Project - Vaho Vishwamitri Abhiyan are examples of innovative ways and their applications. The author's experience and leadership in water management through Gujarat Green Revolution company and other institutions to take up a major 'per drop more crop' initiative which alongwith agri bio inputs showed good speed of adaptation. It is seen that during crisis even small farmers rise to the occasion in adopting new and eco-friendly technologies and convert a major catastrophe into an opportunity. Innovations like GAU - Satellite Krushi Gosthi and 20:20 m-krishi model are examples of ways of reaching the unreached. All these demonstrated and well documented efforts form invaluable ingredients for the coming eco agri revolution.

It is the great merit of this book - ECO AGRI REVOLUTION - Practical Lessons and the Way Ahead' that it shows where and how we may reach the new challenges and shows the way ahead for sustainable developments, a clean environment and equitable livelihood. With much pleasure we welcome this book which will help usher in the Eco Agri Revolution. Eco-agri Revolution is the pathway to an evergreen revolution resulting in enhanced productivity in perpetuity without associated ecological harm.

M.S. Swaminathan

3rd Cross Road, Taramani Institutional Area, Chennai (Madras) -600 113, India
Phone: +91-44-2254 2790, 2254 1698 Fax: +91-44-2254 1319
E-mail: founder@mssrf.res.in, swami@mssrf.res.in

डा. आर. चिदम्बरम्
भारत सरकार के प्रमुख वैज्ञानिक सलाहकार
एवम्
डी.ए.इ.–होमी भाभा प्रोफेसर
Dr. R. Chidambaram
Principlal Scientific Adviser to the Govt. of India
&
DAE-Homi Bhabha Professor

विज्ञान भवन एनेक्सी
मौलाना आजाद मार्ग, नई दिल्ली – 110011
Vigyan Bhawan Annexe
Maulana Azad Road, New Delhi -110011
Tel.:+(91) (11) 23022112
Fax:+(91) (11) 23022113
rajachid@nic.in

Message

Sustainable Agriculture is an essential component of sustainable development. Improved farm productivity should he based on what the author Dr. M.H. Mehta calls eco-friendly and balanced farming. He discusses practical aspects of Eco Agri practices, along with several success stories. The comprehensive discussion in the book includes use of ICT tools in farming, agricultural rehabilitation after natural calamities and his own specialized concept of Bio Shields. In the concluding part of the book dealing with the 'Way Ahead', he discusses how this Eco Agri Revolution can be achieved in practice.

Dr. M. H. Mehta has brought in his rich experience in both industrial research and agricultural practices to write this remarkable and comprehensive book "Eco Agri Revolution - Practical Lessons and the way ahead". I am sure the book would have, indeed should have, a major impact on agricultural planning and practices in India and around the world.

R. Chidambaram

(R. Chidambaram)

9th April, 2017

"Heaven is born in you In the arms of mother earth"

— *Rabindranath Tagore*

Preface

This book is an invitation to think and turn away from the feeling that the fight against poverty, hunger and environmental degradation is too overwhelming, and to start to think of the possibilities and hopes for solving these problems. It is based on the attempts, demonstrated results and search.

I consider myself fortunate that in my journey I got several varied opportunities and rich exposure through leading research institutions, industries, universities, Government and Non-Governments bodies and above all farmer women and men in different parts of the world that taught much of what solutions and the ways I was searching for.

The Sustainable Development Goals, endorsed by the United Nations put great emphasis on moving away from heavy input agricultural system to more eco-friendly and balanced farming. Eco Agriculture or Agro Ecology is the approach capable of producing enough food and accessible food without harming the environment. In the last few decades our awareness and knowledge base for new generation agri bio inputs have considerably increased. At the same time, almost at all world forums, the urgency of converting the lab and factory scale information to actual practices and demonstrated applications have been emphasized again and again.

This book is about moving from 'Know How' to 'Do How' and brings about experience in industrial research and farm scale demonstrations in different parts of the world to demonstrate the practical aspects of Eco Agriculture. As was said by Mahatma Gandhi, "an ounce of practice is better than a ton of advice". Globally we stand in the defining moment in history.The fundamental need in the evolution of happy, peaceful and prosperous nations and societies is laying the foundations for sustainable development. Our Vision and Conviction for the Coming Eco Agri Revolution is vital for a sustainable future.

The book starts with chapters on a holistic view of the world scenario, the interrelationships between man-earth-environment and new technologies and the challenges and emergence of close to nature, eco-friendly technologies.

This is followed by chapters on theory, practices and recent developments in the areas of Biofertilizers, Biopesticides and Biocomposts as important base for Eco Agricultural practices. The global scenario of industrial developments and future trends are discussed based on my involvement, experience and interactions.

In the next few chapters "Do How" of Eco Agricultural practices demonstrations and success stories in different parts of the world are put together. A rich and rewarding experience of working with farmers and farms in different parts of India, Africa and Asia (particularly China,Thailand and Malaysia) is the main base for these chapters.

Encouraged by field studies and success stories at farms, the next part of the book tries to bring out practical lessons and the 20:20 model which emerged as a middle path. It is heartening to see that it is increasingly being accepted as a 'balanced' and practical model. I must acknowledge the interactions with many farmers and their feed back and wisdom all along which made this possible.

I am aware about some repeatations or overlappings but have considered them necessary to maintain continuity and do justice to each chapter.

Throughout my professional and public life in the last 5 decades, working in a spectrum of organizations in varying capacities, I had an opportunity to traverse in different parts of India and the world. We have '600,000 villages in India and eco agricultural model can transform them to a sustainable, strong and happiness centers. The big problem areas can be converted to big opportunity areas.

Promising developments in Water Use Efficiency, ICT, Saline Water Agriculture, Low Cost Green Houses, Bio Shield, Urban Agriculture, Eco Village – Clean Village and even holistic River Revival Projects have relevance and potential to contribute to the coming eco agri revolution. I was considerably fortunate in getting involved or leading these projects and programs and put them in a common thread of eco agriculture.

If the past sixty years led to Green Revolution, the coming decades will be the emergence of Sustainable Green Revolution or Eco Agriculture Revolution. The two goals – Sustainability and Productivity are inter related and should coexist.

I am convinced that in the times ahead, Eco Agri Revolution will play a significant role in our lives and for our future generations. Let this be our common dream and mission. As the great Patanjali has said "when you are inspired by some great purpose, you find yourself in a new and wonderful world".

Acknowledgement

This book is the outcome of many farm visits, presentations, lectures, keynote addresses and interactions in national and international forums. However the interactions with farmer women and men in different parts of India and many other countries have been a major source of learning and I would like to thank all those known and unknown simple folks whose wisdom and feedbacks has made this journey worthwhile.

It is also the time to remember my teams and institutions like I.I.T, Bombay, C.S.M.C.R.I, Bhavnagar, ERL, Arizona & Mexico, B.A.R.C. (Trombay), GSFC Ltd. And GSFC – Science Foundation, erstwhile GAU (All Campuses), CSC – Vadodara. Special thanks are for the young teams of The Science Ashram and Gujarat Life Sciences and the thousands of farmers and field workers associated with them.

Also my special thanks to.........

Professor M.S. Swaminathan for the constant support and guidance and for the inspiring foreword.

Dr. R. Chidambaram, Dr. Li Ji, Mrs. Elaben Bhatt, Dr. Jane Marie Law for the encouraging messages.

Our Global Expert Team, members –Dr. John Fagam, Hans Harren, A Thimmaiah and the new Working Group on Eco. Agriculture – ICFA.

Friends, collaborators and supporters in different parts especially Cornell University, USA and Beijing Agricultural University, China, and Solidaridad Asia Teams.

My Secretary A.K. Joseph for his hard and diligent work.

Team of New India Publishing Agency for urging me to write this book and efficiently taking up the publication work.

And finally to thank my family for their extraordinary understanding and support.

M.H. Mehta
Vadodara
27th April 2017

Acronyms

AAU	Anand Agricultural University, ICFA–Indian Council of Food & Agriculture
BARC	Bhabha Atomic Research Centre
BAU	Beijing Agricultural University
BISAG	Bhaskaracharya Institute of Space Applications – Gandhinagar
BPL	Below Poverty Line
CA	Conventional Agriculture
CIFSRF	Canadian International Food Security Research Fund
CBD	Convention on Biological Diversity
CFS	Committee on Food Security
CIDA	Canadian International Development Agency
CIFE	Central Institute of Fisheries Education
CIMMYT	International Maize and Wheat Improvement Center
C/N	Carbon to Nitrogen Ratio
CSIR	Council of Scientific and Industrial Research
CRZ	Coastal Regulation Zone
CSMCRI	Central Salt and Marine Chemicals Research Institute
CSC - V	Community Science Centre – Vadodara
DBT	Department of Bio Technology
DST	Department of Science and Technology
EA	Eco Agriculture
ERL	Environmental Research Laboratory
FAO	Food and Agriculture Organization
FAI	Fertilizer Association of India
FCI	Food Corporation of India
EFB	Empty Fruit Bunches
FMCT	Fissile Material Cut-off Treaty
FYM	Farm Yard Manure
GAU	Gujarat Agricultural University
GAU SATKRU	GAU Satellite KrushiGosthi
GAPC	Gujarat Agri Processing Co.
GDP	Gross Domestic Product
GGRC	Gujarat Green Revolution Co.
GHG	Greenhouse Gases
GLS	Gujarat Life Sciences
GIAHS	Globally Important Agricultural Heritage Sites
GOG	Government of Gujarat
GOI	Government of India

GNH	Gross National Happiness
GM	Genetically Modified
GSP	Global Social Partnership
GSFC	Gujara State Fertilizers Co. Ltd.
GWP	Global Water Partnership
H-DH	Humidification / Dehumidification Technology
IAASTD	International Assessment of Agricultural Knowledge, Science & Technology for Development
IAEA	Atomic Energy Agency
ICAR	Indian Council of Agricultural Research
ICMR	Indian Council of Medical Research
ICRISAT	International Crops Research for the Semi-Arid Tropics
IAITA	Indian Association of IT in Agriculture
ICT	Information and Communication Technologies
ICFA	Indian Council of Food and Agriculture
IDRC	International Development Research Center of Canada
IFAD	International Fund for Agricultural Development
IGCAR	Indira Gandhi Center for Atomic Research
IIch.E	Indian Institute of Chemical Engineers
IIT	Indian Institute of Technology
ISME	Indian Society of Mangrove Ecosystems
ISRO	Indian Space Research Organization
IUCN	International Union for Conservation of Nature and Natural Resources
IWMI	International Water Management Institute
JAU	Junagadh Agricultural University
KADEC	Kutch Agri Development Consortium
KVK	KrishiVigyan Kendra
LEISA	Low External Input Sustainable Aquaculture
MDGs	Millennium Development Goals
mKrishi	mobile Krishi
MGNREGA	Mahatma Gandhi National Rural Employment Guarantee Program
MoEF	Ministry of Environment and Forests
MSP	Minimum Support Price
MSSRF	M.S. Swaminathan Research Foundation
NABARD	National Bank for Agricultural and Rural Development
NBS	National Bioshield Society
NASSCOM	National Association of Software and Services Companies
NBDB	National Bioresource Development Board
NCF	National Commission on Farmers
NRDC	National Research and Development Corporation
NREGA	National Rural Employment Guarantee Act
NAU	Navasari Agricultural University
NVA	National Virtual Academy
PDS	Public Distribution System
POME	Palm Oil Effluent
RO	Reverse Osmosis

RET	Rare, Endangered and Threatened
RESECO	Remote Sensing and Communications Organization
SEWA	Self Employed Women's Association
SAZ	Special Agricultural Zone
SDG	Sustainable Development Goal
SEED	Science for Equity Empowerment and Development
SEZ	Special Economic Zone
SDAU	SardarKrushinagarDantiwada University
SFAC	Small Farmer's Agri-business Consortium
SHG	Self–Help Group
SICOM	Society of Integrated Coastal Management
SME	Small and Medium Enterprises
SRI	System of Rice Intensification
TSA	The Science Ashram
UNESCO	UN Educational, Scientific and Cultural Organization
UNEP	UN Environmental Protection Agency
UNFPA	UN Population Fund
UNMDG	UN Millennium Development Goals
UPH	Urban and Peri Urban agriculture
USAID	United States Agency for International Development
UA	University of Arizona
UAS	University of Agricultural Science - Dharwad
VKC	Village Knowledge Centre
WFP	World Food Program
WL 1	Wonderlife - 1
WL 2-3	Wonderlife – 2,3
WTO	World Trade Organization

Contents

Agriculture is the perfect example of human endeavor where harmony with nature and earth is absolutely necessary

– *Claude Bourguigun*

1

Man – Earth – Environment Need for New Technologies*

In his classic book 'The Ascent of Man', Bronowski has written, "When I was a young man, we all thought that mastery came from man's domination of his physical environment. Now we have learned that real mastery comes from understanding and moulding the living environment." Our Biosphere, the living environment that supports us all, has problems and we humans are mainly responsible. We do not live 'gently' on the earth. The seven billion of us require air, water, food, energy, housing and amenities and we have developed the earth to provide them and have also created problems. That the problems exist are serious is the bad news. That we are recognizing them and that man has shown ability to organize and grow in ascending order from time to time is the good news.

TECHNOLOGICAL OPTIMISTS OR TECHNOLOGICAL CRITICS

A barrier often divides environmentalists and industry. This is the result of a kind of polarization in the 1970s that forced people to define themselves as either technological optimists or technological critics. The environmental debates have usually advocated environmentally benign, small-scale, relatively "low' technologies. The "soft path" based on conservation and simple renewable energy technologies was pitted against the "hard-path" based on largely on nuclear, fossil fuels etc. In farming Eco agriculture and Organic farming challenged the intensive use of energy, fertilizers, pesticides etc.

WHAT REALLY IS ADVANCED TECHNOLOGY?

For the optimistic proponents of advanced technology, the main danger was that retrealist, anti-technology "prophets of doom" would be causing a collective loss of nerve. To them, it would mean missing an opportunity to get an affluent, supercritical society that would overshoot the limits to growth.

But what exactly is truly advanced technology? Is thee not a misconception in the minds of both the "pessimists" and the "optimists" that Advanced Technology is always Big, Complex and Environmentally Destructive? Before we try to analyze this, let us first try to look at "Earth" as a "Living System".

NATURE'S ADVANCED TECHNOLOGY

Nature itself provides a successful model of an environmentally advanced technology. The biosphere as is now exists is an enormously sophisticated system for gathering and using energy, recycling materials, and regulating the temperature, oxygen level, and many other variable to maintain conditions favourable for life.

Nature was not always so sophisticated. It look billions of years and may "inventions" in the course of biological evolution for the biosphere to reach its present level of "advanced biotechnics". Robert Olson (Futurist, June 1991) and Lovelock (The Ages of Gaia: A Biography of our Living Earth, W.W. Nortan, 1988) have discussed about these "inventions" by the "living earth". Nature solved its original "energy crisis" through one of the greatest innovations in the history of life – photosynthesis. Tapping into the abundant and inexhaustible energy of sunlight, photobacteria, the first photosynthesisers, produced glucose form atmospheric carbon dioxide, replacing the depleted store of organic molecules in the early oceans.

But the invention of photosynthesis had side effects that led to the earth's first "environmental crisis". Photosynthesis produced oxygen as a waste product, and oxygen was highly toxic to fermentation-based organisms. Nature used a temporary "technological fix" organisms called stromalolites, to combine this tree oxygen with iron dissolved in the oceans, creating the iron-rich hematite deposits that we mine as iron ores today. This quick fix only worked for a billion years or so. Then, as the dissolved iron in the oceans was used up, the oxygen level in the atmosphere began to rise sharply. The early biosphere faced a critical situation similar to the one we face today. It could neither tolerate nor recycle its own toxic wastes.

OXYGEN "WASTE" TO ENERGY

Nature responded with a surge of biotechnic innovations. A few type of oxygen-tolerant photobacteria emerged: the blue-green algae. Then another great innovation, respiration, allowed organisms to actually use oxygen "waste" in obtaining energy from organic molecules. The aerobic respirators required much less organic material to sustain them than the earlier anaerobic fermenters, and their yield of energy available for further metabolic processes was 18 times as high.

The fermenters could not compete with the new energy and resource efficient respirators except in oxygen-free environments, such as sediments and deep oceans. As oxygen using photosynthesizers bloomed abundantly over the oceans and land masses, a second planetary environmental crisis occurred that was the opposite of today's greenhouse effect. By burning fossil fuels, we are increasing the amount of carbon dioxide in the atmosphere, which will eventually cause the earth to heat up. The early photosynthesizers used the carbon dioxide in the air and oceans as food and were rapidly eating the blanket that kept the earth warm. The planet was saved from freezing because some bacteria feeding on plant gave off methane. The methane – a far more effective greenhouse than carbon dioxide kept the earth warm. The planet manifested more and more sophisticated biotechnologies that promoted the development of life and protected it from environmental dangers.

ENERGY AND ENVIRONMENT

American environmentalist Ritkin (Entropy, into the Greenhouse World-Bantam Books, NY 1989), has theoretised that we live in a world of decreasing order and resources and an increases in "entropy". An increase in "entropy" means that the system is "running down". But even if we are running down in terms of using up energy and materials, we can minimize pollution with the best possible processes. In this regard, Carl Hodges (Environmental Research Lab – Arizona) argues that whereas industrialization reduces the availability of energy and resources with increasing disorder or imbalance, a biological plant (whether a date tree, a wheat plant or a flower) uses sunshine coming to us from outside the earth to increase order or living balance.

BOX – 1
Holistic Approach

If scientists will have a humane and holistic approach in their fields, humanity would be saved from many hazards and precarious situations which seem to arise only because of partial or fragmentary approach in natural sciences as well as social sciences.

I entirely agree with you that there should be a "SCIENCE ASHRAM" which would be on an international scale and even ultra modern, so that it can attract the notice of and influence the socio-economic trends of global human family.

Whether it is science and technology or it is research in the dimensions of human consciousness i.e. Spirituality, team work is MUST. When the team work is based upon clear understanding and non-subjective motivations of doing Research it would not result in curtailing the psychological freedom of the members of the team.

From a letter from VimlaThakar (A leading Spiritual Thinker) to the author -1992

Our agriculture in many countries of the world has made great progress in terms of productivity – but often at the expenses of balance. It is estimated that American farmers use as much as 10 calories of fuel energy to produce 1 calorie of food energy. (As against this, the traditional dry land Asian rice culture uses about 1 calorie of energy (human labour) to produce 20 calories of rice).

The world, in the future, must have a productive "ordering" or "balanced" biological activity. To provide such a system will require application of all sciences, even the development of new sciences and great cooperation between individuals, organizations and nations. Peter Drucker – the doyen of management sciences calls for the requirement for transnational ecology.

We will therefore have to remember that whereas reckless use of chemicals either industrial or agricultural has to be opposed, we must at the same time recognize that we need alternatives to produce and protect our food, fiber and timber. We will need advanced technologies for ever increasing demand and we must remember that it is a misconception that advanced technology is not environment friendly. Technology that is intelligently conceived and carefully implemented for agriculture and industry, including measured and careful application of fertilizer and crop protection chemicals, represents the hope for the future and, as such, should be a cornerstone for agriculture and industrial policy for the coming decades.

ENVIRONMENTALLY ADVANCED TECHNOLOGY

A technology shaped by environmental values and drawing on the full potential of modern science will eventually be different from a technology guided only by value-empty economics – but it will still be an advanced technology.

An environmentally advanced technology will have several characteristics. Firstly, it will emulate the characteristics of nature itself. It is possible to list the main characteristics of environmentally advanced technology as follows :

1. **Sustainable:** Capable of satisfying present needs without jeopardizing the prospects of future generations.
2. Based on safe and exhaustible supply of energy. Just as nature met its energy needs by "inventing" photosynthesis, invention of efficient use of sunlight, safe and inexpensive method for fusion etc.
3. High efficiency use of energy and other resources, improved production processes, buildings, transportation etc.
4. **High efficiency recycling and the use of by products:** we would always use effluent of one process as a raw material for other.
5. **Intelligent:** Just as biosphere evolved increasingly intelligent life-forms, our tools are evolving to a higher level of information – processing capability.

6. **Increasingly alive:** There will be emergence of technology "living machine" – composed by many types of organisms and advanced structural materials designed to perform vital functions like food and fuel production, waste treatment, water purification.

ECO AGRICULTURE – AGRO ECOLOGY

The sustainable developmentalgoals of United Nations greatly emphasizes the need to transform our current input of heavy food system to more sustainable system. In a country like India where majority of the population still depends on agriculture farming by eco agricultural methods is the way forward. Eco agricultural practice seems to be the best way to address the challenges like Hunger, Malnutrition, Poverty, Environmental Degradation and Climate Change. Eco Agriculture is still evolving among conventional forms of Agriculture and the developments of new generation bio inputs like Bio Fertilizers, Bio Pesticides and conversion of far agro waste to Bio Composts will play the most significant role in the coming years. A few years back, our team of young scientists of the NGO, The Science Ashram while interacting with farmers met with some unexpected feed back. In many of the areas where farmers using heavy doses of Pesticides and Fertilizers were ready to listen to the talks about the harmful effects and the decreasing net income and environmental problems. However, they kept asking – "are there alternatives?" When explained that the new generation agri bio inputs, their simple reaction - "you make them and give to us - we want to try". This compelled the team to start a production unit - Gujarat Life Sciences to make complete packages of agri bio inputs. This is a typical example of the emerging trend in many parts of the world.

BOX – 2

Agro Ecology / Eco Agriculture

Agro Ecology mode of agricultural development has shown notable success in the last decade. It was created by the convergence of two scientific disciplines : agronomy and ecology. It seeks ways to enhance agricultural systems by mimicking natural processes, thus creating beneficial biological integrated and synergies between components of agro system. Agro ecology is a coherent concept of developing future farming systems.

It has proven potential to significantly improve yields, reduce rural poverty and better sustainability.

From report on Right to Food Oliver Schutter – Human Rights Council – UN A/HRC/16 49 December 2010

Economists see the environment as a subset of the economy, whereas ecologists see the environment as a subset of the environment. Lester Brown discusses the concept of the "Eco – Economy" in the beautiful book "The Earth from Air" (Yann Arthus – Bertrand) based upon the real costs that result from the degradation of the environment and the development reconciling economy with ecology. An economy is sustainable only if it respects the principles of ecology. Nature relies on balances which include balances between soil erosion and new soil formation, carbon emissions and carbon fixation and trees dying and trees regenerating. An Eco – Economy will give us a world where we are a part of nature, instead of 'apart' from it.

Building a new economy means phasing out old industries and creation of new, innovative, eco-friendly enterprises. These will create new innovations, opportunities, new jobs and structural changes in the global economy."Green

BOX - 3

Eco Agriculture

The term "Eco Agriculture" was coined by Charles Wattens (founder of Acres Magazine) in 1970 to unify under one umbrella the concept of "Ecology" and "Economical" with belief that unless agriculture is ecological it could not be economical.

Eco Agriculture is both a Conservation strategy and a rural development strategy. Eco Agriculture recognize agriculture produces and communities as key stewards of eco system and bio-diversity. It implies an integrated ecosystem approach to agri land scales to address the 3 pillars.

1)Conserving Bio-diversity

2)Enhancing agricultural production

3)Improving livelihoods

Farmers as ecosystem partners.

Contribution of Eco Agriculture for Millennium Development Goals (MDG) (now Sustainable Development Goals - SDG) will not be reached without securing the ability of the rural poor to feed their families and gain because, while at the same time protecting the bio-diversity and ecosystem of the estimated 800 million people who do not have access to sufficient food, half are small holder farmers, one fifth are rural land less and one tenth are dependent on rural land, forests and fisheries. For most of them, reducing poverty and hunger will depend mainly on their ability to sustain and increase crop, livestock, forest and fishery production.

A key opportunity for enhancing progress towards the MDGs is instrument in locally driven land management approaches such as eco agriculture strategy that built upon synergies between rural livelihoods, enviro sustainability and food security.

(Ref.: Acres Magazine)

technology" will provide a vast range of services – from better ways of growing, breaking down of toxic wastes or brand new process routes which do not generate any waste streams. It is believed that whereas "gray" technologies will not disappear all together, the green technologies will become dominant in 3 or 4 decades.

CONCLUSION

The purpose of this discussion has been to try to take a holistic view of man - earth and environment, their interlinks considering earth as a "living" system and the optimism in "man" and his ability to create new technologies. In a situation of rapidly worsening environmental impacts on a global scale and growing understanding that the success of human civilization is at stake, the social and political pressure of creating environmentally advanced technologies will become enormous.

We are all afraid for the future, for the world. That is the nature of human imagination. And through every crisis in history, man has gone forward. The message is clear – problem exists – problem is recognized – Man is capable. Technological optimism has a strong presence and should be used to lead us in the 21st Century.

REFERENCES

1. Jerome G. Glenn, Future Mind, Acropolis Books Ltd., Washington D.C. 1989.
2. Mehta M.H., "Man – Earth – Environment", Lecture given at Japan Creativity Society Tokyo, March 1990.
3. Lovelook James, The Ages of Gia, Oxford University Press, 1988.
4. Personal discussion, Rev. VimlaThakar, Mt. Abu, 1989 – 1992.
5. IAAST – International Assessment of Agricultural Knowledge, Science and Technology for Development – Report, 2009.

"Remember your humanity, and forget the rest. If you can do so, the way is open to paradise; if you cannot, there lies before you the risk of universal death"

– Bernard Russel & Albert Einstien
In their famous joint manifesto in July 1955

2

Science with Compassion for A Hunger-Free World

It has been repeatedly proving that Science& Technology that is close and in harmony with nature is proving winner in the long run. If the past 60 years led to the Green Revolution in agriculture which often became Greed Revolution, the next 60 year should be the emergence of love for nature, diversity and Sustainable Green Revolution. The two goals – Sustainability and Productivity are inter-related.

It is also said that the main difference between the past and today is that our problems are truly global. We are also aware that globally and nationally the rich-poor divide is growing and though world produces enough food, nearly 800 million people remain food insecure. Among 8 millennium goals, eradication of poverty and hunger and environmental sustainability are perhaps the most prominent and complimentary.

In his book, "Remember your Humanity – Pathway to food Security", M.S.Swaminathan has emphasized that a pro-nature, pro-woman and pro-livelihood orientation should be the bottom line of all technological development and delivery program for food security. Mainstreaming of environmental concern in agriculture to avoid exploitative agriculture and ecological consequence is therefore the call of the time.

COMPASSION

The great Tibetan scholar and close companion of HH Dalai Lama, Thupten Jinpa defines compassion as a mental state endowed with a sense of concern for the suffering of others and aspirations to see that suffering is relieved. One of the highest understanding underlined at the global level and in the preparation of sustainable development goals is for a hunger free world and for that helping the needy farmers in the third world.

For the search of a sustainable future, most people now talk about humanism, wisdom and compassion. The major difference between optimists and pessimists is their position in the debate whether human beings are able to operate collectively and in cooperation on the basis of love and compassion. The pessimism of some people is now regarded as the main cause of unsustainability.

The Sustainable Development Goals (SDGs) replace the eight Millennium Development Goals (MDGs), which expired at the end of 2015. The MDGs were established in 2000 following the Millennium Summit of the United Nations. Adopted by the 189 United Nations member states at the time and more than twenty international organizations, these goals were advanced to help achieve the following sustainable development standards by 2015.

1. To eradicate extreme poverty and hunger
2. To achieve universal primary education
3. To promote gender equality and empower women
4. To reduce child mortality
5. To improve maternal health
6. To combat HIV/AIDS, malaria, and other diseases
7. To ensure environmental sustainability (one of the targets in this goal focuses on increasing sustainable access to safe drinking water and basic sanitation)
8. To develop a global partnership for development

While talking about the next revolution : Sustainability, in "Beyond The Limits", Meadow et al have pointed out :"of course no one knows how to bring about a sustainability revolution. There is not and will never be a checkoff list like "To accomplish a global revolution, follow the twenty steps below". Like the revolutions that came before, this one cannot be planned or dictated. It would not follow a list of fiats from a government or from computer modelers. The Sustainability Revolution, if it happens, will be organic and evolutionary. It will arise from the visions, insights, experiments, and actions of billions of people".

THE ALTERNATIVE MODEL

Although chemical agriculture has till recently been considered "modern agriculture", it is now recognized that eco agriculture are at the leading edge of scientific farming. For brevity, we can use "eco agriculture", "agro ecology" and "organic farming" in a broad sense. There are many other synonymous terms such as "regenerative agriculture" etc which all describe agricultural practices or systems that largely overlap with "eco agriculture and organic". This is elaborated in Box – 1.

BOX – 1

ALTERNATIVE OF CHEMICAL AGRICULTURE

Although chemical agriculture has, until recently, been considered "modern agriculture," it is now recognized that organic agriculture and agroecology are at the leading edge of scientific farming. Organic agriculture and agroecology employ the same farming techniques and technologies. Both use no synthetic inputs, such as chemical fertilizers and pesticides. Instead, fertility, pests and diseases are managed using a combination of sophisticated practices and inputs that make use of the most advanced scientific knowledge relevant to agriculture, including entomology, agronomy, plant physiology, soil biology and chemistry, plant genetics, meteorology etc. to manage fertility, pests and diseases and maximize quality, nutritional value and resilience of the resulting crop. Importantly both strive for self-sufficient, on-farm or local production of all fertility and pest management inputs from renewable on-farm resources and have a strong focus on soil health.

Organic and agroecology differ, not in the farming methods used, but only in the fact that organic farms undergo an independent verification process that independently verifies their exclusive use of the practices described above. This independent verification adds value to the product, improving the financial returns of the farmer.

For brevity, we will use "organic agriculture", "organic farming", or "organic" inclusively to indicate both "organic agriculture" and "agroecology". There are many other synonymous terms, such as, eco-agriculture, regenerative agriculture, biodynamic agriculture, etc., which describe agricultural practices or systems that largely overlap with organic and agroecology.

Ref: J. Fagan, H.R. Harren, M.H. Mehta and A. Thimmaiah Concept Note "Scaling Up Organic Agriculture and Agroecology", February 22, 2016

Farmers worldwide are greatly concerned about the high input costs for fertilizers, pesticides, seeds, etc. Indian farmers spend more than 20 lac million Rupees on these inputs, in spite of the fact that more than Rs.5 lac million goes for fertilizer subsidy. Fertilizer use efficiency is drastically declining for the reasons discussed earlier. As regards pesticides, the scenario is changing drastically. The European Council last year passed a directive banning more than 40% chemical pesticides and many Government (like Sweden) have started making a blacklist of a number of toxic herbicides, fungicides and insecticides. This, coupled with new advancements in bio pesticide industries, technology of organic movement is already bringing a major change. The European Parliament plenary will vote on proposed legislation to subject bio pesticides to a quicker approval process than conventional pesticides.

GLOBAL TRANSITION TO SUSTAINABILITY AND NEW AGRICULTURAL REVOLUTION

Global economic output increased 22 fold during the 20th century as the world population increased four fold. Most people are healthier and live longer and yet almost a billion people go hungry. Our world is one of deepening inequality.

Also, we are the first generation to recognize that humanity has become a force to undermine the Earth's capacity to support our future. In the Anthropocene, we are in the driver's seat. We have the creativity and innovativeness to bring in global sustainability.

BOX – 2

As a Scientist, I firmly believe that Science shares a responsibility to push for action and solutions rather than simply study and document.

Prof. Yuan Lee
1986 Nobel Laureate

We need to develop a new global leadership for sustainable development. An important starting point is to create a deeper understanding of our society.

G.H. Brunderand
Former PM of Norway

A global transition to sustainable agriculture while securing food for all is our aim. World agriculture is today the largest cause of bio diversity loss and green house gas emission (nearly 30 percent of GHG) as explained by Rockstream& Klum in the beautiful publication The Human Quest. They further argue that we must break the current dischotomy between conventional and ecological agriculture, which has mostly led to unconstructive wars between different experts and practitioners. "Instead, we should admit that neither the conventional practices using commercial fertilizers and pesticides or only organic method can meet the requirements of staying within safe planetary boundaries". We must tap the best ideas, wise practices and scientific break through to shape new generation of sustainable farming'. For that a holistic approach is the need.

Working with the earth quake affected farmers in Kutch – North Gujarat, where rehabilitation through agriculture was a priority, our team searched most appropriate model. An Integrated model involving agri horticulture and livestock turned out to be the best answer for empowering the local farmers. It was realized that integrated farming alongwith improved water management goes a long way in big and sustainable improvement in productivity.

When we compare such models with "Mega Farming" model – the overall change to human societies and environment clearly show the contrast. Philip Lynbery in his book "Eat less meat or else the planet will starve" argues strongly against 'Mega Farming" from the point of view of compassion and planet health. As he says, "there is no end to techno-idiocy in the pursuit of profit. Hacking down of South American forest to grow feed pigs and Chickens of China is wasteful and inhuman approach. Further the effluent from mega farms of cows and pigs across the world poison river systems and water bodies. The cereal crops now fed to animals, largely cattles, could feed nearly 3 billion people – a simple solution to the problem of expanding world population"

From the point of view of an average farmer, the priority is on a model, which gives good productivity and reduces input costs. Technology, which is sustainable and of long term benefit will be intelligently accepted by a farmer and his judgement is usually the final test. It is thus time to view such matters from the practical field conditions and take actions.

CONCLUSIONS

1. "Each 1 percent increase in agricultural productivity reduces poverty by 0.6 percent'. A yield increase of 1 percent decreases the number of people living with less than US$ 1/day by 6 million. Agricultural progress represents the best safety net against hunger and the greatest good it confers on the largest number of people.
2. Ecologically improved technologies can lead to higher productivity without adverse impact on the natural resources and environment. Blending traditional and frontier technologies lead to give combined strength to Economics, Energy, Equity and Sustainability.
3. Translating ecofriendly technologies into action means improving productivity of small (and big) farmers and well-being of people. One of the most important needs of the world is developing a model to improve farm productivity at lower input costs in a sustainable manner.
4. A hunger-free and nutrition secure world is possible. There is a vast untapped potential. We can support all the world's people adequately and sustainably long into the future. We need to act and there are ways.

In our recent meeting with HH Dalai Lama at Dharamshala, he laughingly said, "if the Buddha Shankyamuni were to return among us he would be an ecologist".

The great power of science can benefit the world when we know to temper it with compassion and love. As said by Vivekanand : "We want today the bright sun of intellectuality joined with the heart of Budha."

You must be the change you wish to see in the world.

– *Mahatma Gandhi*

3

The Coming Eco Agri Revolution – Why India Should Lead?

The words "Economy" and "Ecology" have the same root. Both these words are derived from Greek 'Oikos' means house or house holds and both imply thrifty economical way of dealing with valuable resources. However, for some reason we seem to find these two words coming opposite each-other in many situations.

The destruction and degradation of the environment has become the greatest problem of our century. Our survival will depend on how well we are able to redefine our technologies and our relationship with nature. We therefore need a new sustainable eco-agri revolution and a country like India is in an ideal situation to give a model for that. In the new economy, good ecology will be fundamental to good business.

BEYOND GREEN REVOLUTION

We have often emphasized that perhaps the two greatest successes in India since independence have been maintaining a stable democracy and success in bringing the green revolution, which changed us from a food-begging country to a food self-sufficient nation. We should rightfully feel happy and proud about these achievements, but at the same time remember that the global problems, environmental issues, priorities and the technological scenario have vastly changed in the last two decades.

Another feature of Indian agriculture is the vast varieties of agricultural and horticulture practices as shown in Fig. 1.

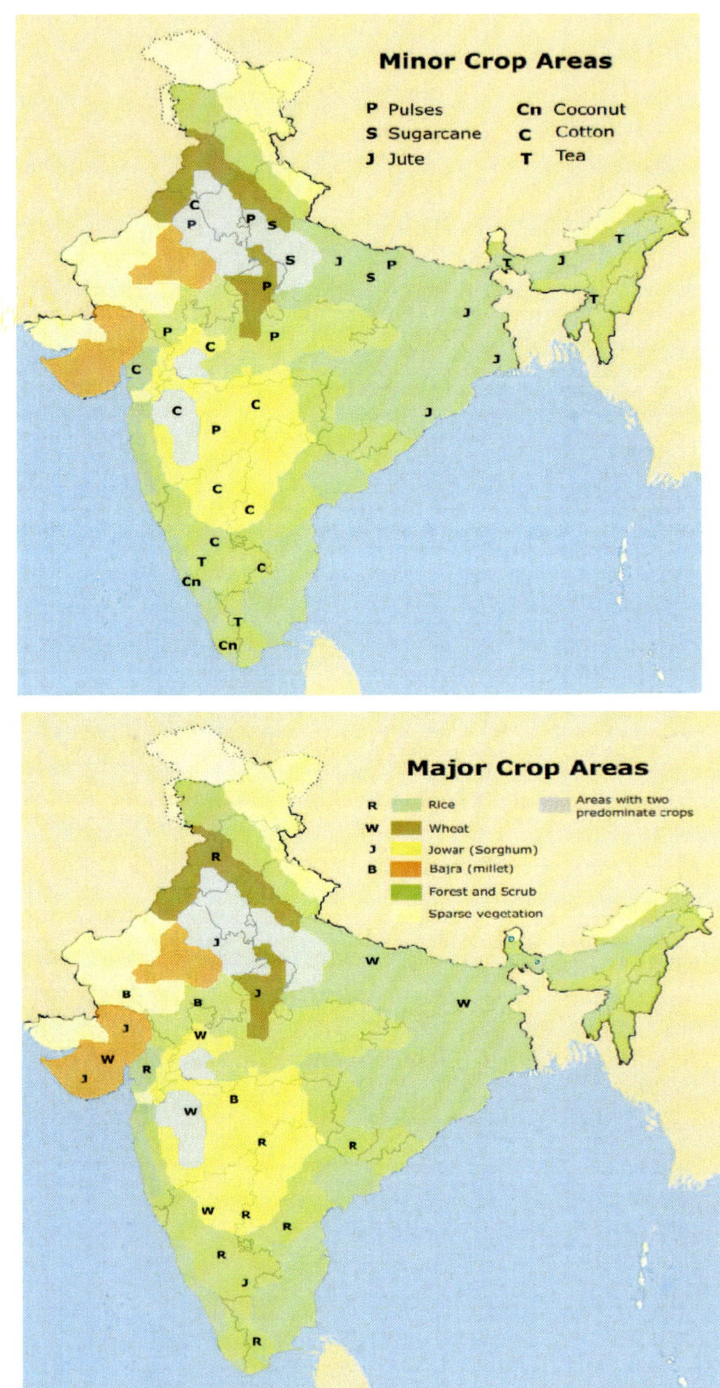

Fig. 1: Minor and Major Crop Areas of India

In the picture today, the following have strikingly emerged:

(a) The farm productivity have been going down particularly in the green revolution areas like Punjab, Haryana. Closely linked with this is the drastic reduction of carbon in the soil and lowering of the fertilizer use efficiency.

(b) The problem of contamination of chemical pesticide in food, milk, river and water sources have been alarming in many situations.

(c) The worldwide awareness of harmful effects of organo-chlorine and organo- phosphorous pesticide is clearly making impact on the Indian chemical pesticide industry.

(d) Organic movements are on the rise.

(e) Rain-fed agriculture is being increasingly looked up as a major sector to shoulder the larger burden of providing foods in developing countries.

(f) Biopesticide industries are rising at a very high rate, though its size and contribution on a global scale is much smaller than chemical pesticides.

There is thus the time and need for the next revolution- Eco Agri Revolution – A revolution that is sustainable, balanced and farmer-friendly.

INDIAN SCENARIO

Figure 1 gives idea about the diversity of crops and environmental conditions in India. This together with the fact that India has the largest number of small farmers and a history of sustainable farming practice makes it a typical place for Eco agriculture revolution. Some of the features of Indian agriculture are given in Box 1 and details of share of small and marginal farmers in land possessed and in crop output are given in Box 2.

BOX – 1

Indian Agriculture

- India is the second largest producer of agri products and account for 7.68% of total global agri output.
- Agriculture contributes 17% to GDP of India's economy.
- At 190 million ha, India has the 2nd largest agri land holding in the world.
- Around 800 million population depends on agriculture as primary source of income and livelihood.
- Agriculture in India provide employment to 56% of the Indian workforce.
- Indian agriculture is the home of small and marginal farmers (80%). The future of sustainable growth and food security in India depends on the performance of small and marginal farmers.

Contd.

- Average size of farms in India is 1.37 ha.
- The role of small farmers in poverty reduction is well recognized.
- The fertilizer use /ha is inversely related to farm size both irrigated and unirrigated farms.
- GDP growth originating in agriculture is twice as effective in reducing poverty as GDP growth originating outside agriculture. Small holdings plays important role.
- Efficiency and productivity point of view, small holdings are equal or better than large holdings
- The fertilizer use /ha is inversely related to farm size both irrigate and unirrigated farms.
- Efficiency and productivity point of view, small holdings are equal or better than large holding.

Share of small and marginal farmers in India - Rishi Nair – Agro News, Nov. 2016

BOX - 2

Share of Small and Marginal farmers in land possessed and in crop output (%)

States	Share in Area (Land Possessed)	Share in Output
West Bengal	83.0	86.2
Uttaranchal	81.4	46.7
Kerala	79.8	78.1
Himachal Pradesh	73.7	73.9
Orissa	73.2	72.5
Jharkhand	73.1	78.1
Assam	72.4	70.7
Bihar	67.2	69.2
Jammu and Kshmir	66.4	77.3
Uttar Pradesh	59.4	65.1
Tamil Nadu	54.4	51.7
Chattisgarh	49.4	46.1
Andhra Pradesh	47.5	46.7
Haryana	42.2	29.9
Karnataka	37.4	38.5
Madhya Pradesh	34.3	27.9
Gujarat	34.3	35.3
Maharashtra	31.7	35.2
Punjab	29.9	19.3
Rajasthan	22.6	33.2
All India	46.1	51.2

Small farmers in India - S. Mahendra Dev. – Indira Gandhi Inst. Of Development & Research, Mumbai, June 2012.

REGENERATING THE SOIL

By losing their organic matter and biological activity soils are becoming the most limiting factor of agricultural production. When soils lack fauna and microbes, they become compact and fragile and susceptible to climatic attacks. Organic matter does not decompose if there are no microorganisms. Methyl bromide used for soil fumigation was found to kill microbes up to 25 meter depth and restoring it would take several years.

Agriculture is the perfect example of a human endeavour where harmony with nature and earth is absolutely necessary. Soil formation is the result of a synergy between clays from mother rock and humus from the organic debris.

Decline in soil organic carbon content (SOM) of soil in states like Punjab, Haryana and many others is due to continuous cultivation of cereal based cropping system is clearly a big problem. The SOM is extremely important for crop productivity and particularly for poor soil of arid, semi-arid and coastal areas. African countries are losing fertility at an alarming rate. To develop a sustainable agriculture for the crops and small lands must develop agricultural practices which can maintain high level soil biological activity and deep route system. SOM plays a major role in the retention of cationic nutrient by dominant soils of these areas with weak ability to help nutrient cationic.

AGRO WASTES UTILIZATION & BIOCOMPOSTING

India generates more than 5 million tons agro wastes every year. Some agro wastes finds use as fodder, fuel and a large amount is burnt wasting a good resource and damaging the soil and environment. The largest fraction of all organic carbon entering the soil is that contributed by soil residues.

Microbes are the garbage disposal agents of nature. Use of products like Wonderlife 1-2-3 and Relife can breakdown the agro wastes and convert them in to useful Biocompost. Such products are a consortia of efficient micro organisms and they decompose the agro wastes and convert them to useful source of soil nutrients. A large scale programme of Biocomposting will thus be the first step for Eco agriculture.

BIOFERTILIZERS

In the last few years, our appreciation and understanding of the role of Biofertilizers has considerably improved. Table-1 lists some of the Biofertilizers along with other agri bio inputs (ABI). The activities of microbial and animal species- including bacteria, algae, fungi, mites, millitodes and worms condition soils, break down organic matter and release essential nutrients to plants.

Table 1: Agri Bio Inputs (ABI) – Typical Range

Nutrients	Biofertilizers : Microbial	N - fixation P - solubilisation S - solubilisation
	- Biofertilizers cum Biofungicides	
	- Agro waste de-composers	
	- Enrich Biocomposts	
Pest Control	- Microbial Biopesticides e.g. Trichoderma, Bt, Bs, BB, etc.*	
	- Botanical Extracts - Neem, Karanj, etc.	
	- Pheromones	
Growth Enhancers	- Seaweed extracts	
	- Amino acids	
	- Biologicals	

*BS - Baccilus Sparieus

*BB - Beveria Basiana

From the earlier time of development of monocultures of Rhizobium, Azotobacter, Acetobacter etc. now development of multi-microbial consortia are proving more cost-effective and farmer-friendly. In a large number of applications, Biofertilizers have been able to cut down the requirement of urea and DAP up to 20%. In rain-fed areas, their role will be even more important.

BIOPESTICIDES

Biopesticides have drawn much attention in the last couple of decades as big alternative hope. The concept of biopesticide has often been described in the following manner (Reference : Marrone Bio Innovations, CA-USA).

Biopesticides or natural pesticides are reduced-risk products derived or developed from biological or naturally derived chemistry. Biopesticides offer value to users by providing a combination of both effective performance and product safety. Biopesticides perform efficaciously while providing the flexibility of minimum application restrictions, superior residue resistance management potential, and human and environmental safety benefits.

SOME OF THE UNIQUE FEATURES AND BENEFITS OF BIOPESTICIDES INCLUDE

1. The ability to provide alternative modes of action to traditional products which makes them a critical component in most IPM programs;
2. Registration in less time than conventional chemical products because biopesticides exhibit minimal environment and humans;
3. The ability to extend the life of conventional chemicals by providing resistance management benefit programs;

4 Exemption from tolerances, such as reduced pre-harvest restrictions and application in environment which permits biopesticides that have no Maximum Residue Levels (MRLs) to be used on crops in urban settings;

5. A high degree of worker safety and the shortest recently intervals allowed by law;

6. Value-added benefits, such as improved plant health, yields and quality and an increase in beneficial and organic cultivation programs.

A comparison of biological v/s chemical pesticide gives a good idea about the greater need for biopesticides and bio-control agents and is discussed in details in the chapter on Bio Pesticides later.

GLOBAL BIOPESTICIDE INDUSTRY

The global market for biopesticide was $ 1.6 Billion in 2009 and is expected to be $ 3.3 Billion in 2014 i.e. 15.6% growth rate as per the study by BBC Research of USA. That growth rate will be much higher than the expected rate of 3% for the synthetic biopesticide industry. However the volume of chemical pesticide industry is several times larger than biopesticide industry.

The second important feature of biopesticide industry is that mostly it is in the SME segment. This is a good news for a balanced economic growth for the world. It also emphasizes tremendous challenge for R&D and new development as also organizing the industry for registration and global marketing network. One likely trend will be international cooperation and networking for creating win-win situation.

The estimated growth rate as mentioned above can significantly change in the coming year when we look at major sectors in the future:

i) Organic farming

ii) Rain-fed Agriculture

Biopesticides which are certified by international certifying agency like Ecocert and IFOAM, etc. can have a substantial contribution in these two areas in the time to come.

CHEMICAL V/S ORGANIC

Is the world vertically split in the issues of ecology and agriculture, especially organic farming? Recently Amir Khan's 'Satyamev Jayate' TV episode has reached the largest number of people on the issue of chemical pesticides. This was one of the fastest extension work on the subject!

We often get two extreme points during such debates : One- Take away chemical fertilizers and pesticides and half the world will starve; Second- Organic farming can feed the world at lower cost and in a sustainable manner.

As against the claims and counter claims, the scenario today is worth examining.

i) The farm productivity has been going down particularly in the Green Revolution places like Punjab, Haryana, *etc.* Closely linked with this is the drastic reduction of organic carbon in the soil, which has direct effect on the lowering of the fertilizer use efficiency.

ii) The problem of contamination of chemical pesticides in food, milk, river and water resources have been alarming in a large number of situations. A report published in Down to Earth – February-2011 under the title "Pesticide Rich Food" had listed food items from Cauliflower, poultry, milk and butter to apple and tomato, which are highly objectionable to life as for pesticide residues. Each of these pesticide residues had well known negative health impact including on nerve system, liver, kidney, lungs, reproductive system, *etc.*

iii) The worldwide awareness on such a harmful effects have resulted in banning of more and more chemical pesticides. In a historic development, the European Parliament black-listed more than 50% of prevailing chemical pesticides.

iv) The use of *Bt.* Cotton has brought down the chemical pesticide consumption substantially (cotton has been the largest user of insecticides).

v) Rain-fed agriculture is being increasingly looked up as a major sector to shoulder the large burden of providing food in the developed countries. This system traditionally uses much less or no agri inputs.

vi) Organic awareness and organic movement are on the rise worldwide.

vii) Similarly the bio-pesticide industries are rising at a very high rate.

viii) Organic farming will be possible if crop-livestock farming systems can be adopted. In India, nearly 80% of farmers possess less than 1 ha land. For such farmers, organic or green farming will be the best alternative.

ix) Only 30% of total cultivable land in the country is covered with chemical fertilizers (where irrigation facilities are available) and the remaining 70% of arable land, which is mainly rain-fed and use no or negligible amount of fertilizers.

BIHAR TO BIKANER – CASE STUDIES

In a majority of situations involving small farmers in backward or poor district, the problem has shown to be converted in to advantage. Such was the case of Bihar-Nalanda district, where thanks to the State Government's initiative on organic farming project, was promoted and a package of agri bio inputs : Biocomposts, Biofertilizers & Biopesticides, such a Wonderlife-G, Tricholife, Biocompost, Neem-A-Life, *etc*. This package was standard and prepared by joint teams of State Agriculture Department, Gujarat Life Sciences and Ecocert. The local farmers adopted it so well that it proved to be a boon and a World Record in Potato production was reported. Knowing that a majority of the farmers in developing world (they are the most important people for a hunger-free world), are in such a situation- a model like this may be a better solution.

Another such example has been for the organic farming of Psyllium (Isabgul) in Bikaner- Rajasthan. Psyllium is an important and premium crop in North Gujarat and Rajasthan. A large number of marginal farmers in this area have been organically growing this crop and export through an agency in Europe. The yields were low and no agri inputs were used. Use of Eco-certified Biofertilizer increased the yield substantially (up to 30%) and the chances of failures considerably decreased due to improved soil, moisture conservation and nutrients. The farmers' income increased and the exporter was doubly happy! There are similar success stories in other parts of the world. What is important is that with low cost agri bio inputs, considerable improvement in production is being demonstrated.

20:20 MODEL THE MIDDLE PATH

In spite of several interesting developments in the area of biopesticides, we cannot wish away the fact that nearly at $ 2 billion, biopesticide industry is extremely small as compared to the $ 45 billion chemical pesticide industry. The second argument often is against the biopesticide industry that it does not have solutions for all kinds of plant diseases and insects. Unfortunately even our agricultural universities and research institutions are not fully aware about the alternatives and mostly make recommendations for conventional chemical pesticides year after year even when the alternatives are available. Similarly the giant fertilizer industry will be ever eager to play its role on agriculture but will keep appreciating the role of ABI with increasing respect.

On the other hand, biopesticides are being looked at with great hope for improving environment and farm productivity and their growth rate have been much higher than chemical pesticide industry. But much more is required to be done.

A new and applicable model for most areas has been emerging as a practical and balance one, which we have named as 20-20 Model. This model basically emphasizes higher farm production with lower input costs by judicious use of the available developments of agri bio inputs and doing away with the overuse of chemical inputs or minimizing it to a great extent. Such a 20-20 Model has been demonstrated for different crops and under different conditions in India, especially Gujarat, Tamil Nadu, Bihar, Rajasthan, etc. It is also now being implemented in some of the African countries after field demonstrations and actual use by the farmers. Again both for rain-fed or organic farming areas, the complete packages of certified agri bio inputs have been giving excellent results. This indicates a great opportunity for a number of marginal and poor farmers. Making them popular and reaching the unreached would be a major challenge land need of the time for a sustainable and balanced growth. Locally there are several groups, NGOs, companies and institutions, who are working with missionary zeal and as such, such a noble and dedicated efforts have started showing results. A paradigm shift in the government policies is required now with respect to special subsidies, massive programmes and special fund for R&D and Demonstration. A massive support for the new generation agri bio inputs and concentrated efforts can make India a leader in giving a new ecofriendly agriculture model to the world.

We need food justice and food security as a basic human right through fairer distribution of food as a means of ending chronic hunger and malnutrition. Eco Agri Revolution should take us to the millennium goals of eliminating hunger, poverty and improving the environment. With the largest number of small farmers in the world, with its vast varieties of agriculture, horticulture and integrated farming and the remarkable capacity of the farmers to adopt new and eco friendly products and practices – India is in an ideal situation to lead the world for eco agriculture revolution.

REFERENCES

1. Claude Bourguignon, 'Regenerating the Soil' Other India Press, 2005.
2. Swaminathan M.S., 'Science and Our Agricultural Future' Inter Disci. Dialogue on Biotech Chennai, Jan. 9-12, 2003.
3. Mehta M.H. "Biopesticide Industries- Global Challenges and Opportunities", Presidential Address, 6th Int. Conf. on Biopesticides, Chiangmai, Thailand, Dec-2011.
4. Jayraj S. The Hindu Survey 2002.
5. Arjun Singh Mehta, et al – Scientific Organic Farming Package for Bihar, National Seminar Patna, 22-23 June, 2011.
6. Mehta M.H., The 20-20 Model, Agri Year Book, 2009.
7. World Record in Potato Production – Organic Way - Hindustan Times, Patna – March 15 2012.

You have feet. Why do you pretend to be lame? You have hands. Why do you hide your fingers?

– Jalaluddin Rumi

4

Sustainable Agri Biotech Revolution for Africa

African agriculture is in crisis, and Africa's farmland is losing its fertility at an alarming rate. Farmers grow the same crops in the same fields year. They don't fertilize, and they don't terrace their fields. When the soil wears out, the clear a new piece of land (wwww.npr.org). N. Kozana Diamini Luma, the Chairperson of the African Union Commission has emphasized that 'the year 2014 – the year of Agriculture in Africa' will be used as an opportunity to accelerate the drive for food security. The major thrust for that has to be on ensuring that the farmers make their profession on economic activity that generate wellbeing for all in a sustainable manner [1, 2].

Africa will have a population of 2 billion people by 2050, the majority women and youth. This prediction and the climate change problem summarize the scale of agricultural challenge for them. Currently more than half of all people living in Africa depend on agriculture for all or part of their livelihood.

EVER GREEN REVOLUTION OR ECO AGRI REVOLUTION FOR AFRICA

Nearly all agree that Africa needs technology for farming comparable to the Green Revolution of Asia. In 2006, two major philanthropic foundations namely Gates Foundations & Rockefeller foundations joined in a joint venture called Alliance for a Green Revolution in Africa (AGRA). They emphasized the importance of introducing African small fodder farmers to modern agricultural science. However, this was rejected strongly by some calling it "naivete" about the causes of hunger and warned that this approach of "Industrial Green Revolution" would result in worsening exclusion and marginalization of African country side 2.

Though developments in the form of high yield "Green Revolution" seeds helped bring hundreds of destitute farmers in Asia out of poverty, these technologies had only limited application in Africa, because they require high use of chemical fertilizer and irrigation. These are not easily available or affordable for most African small holders.

The report of international Assessment of Agricultural Knowledge, Science & Technology for Development (IAASTD) under the auspices of the World Bank & The U.N. is widely cited 3 and lays emphasis on agro ecological approach.The best approach is usually to combine agro ecological insights with Green Revolution seeds plus agri-inputs(3). For example, Integrated Pest Management with Bio Pesticides, Bio Fertilizers etc. This, then became our important mission while working with farmers in different parts of Africa about which some more discussions and case studies will follow this chapter.

SMALL FARM HOLDERS & BIG HOPE

I have observed during my visits to different rural communities in Africa that most of them are small farmers and poverty and hunger problems trace directly to the low productivity of labor in the growing of crops and the herding of livestock. Further, nearly 75 percent of all farm labor is performed by women. They are busy, efficient but have too many things to do and are in a "poverty trap" due to lack of productive technology.This has led them to use less sustainable methods as shown in Box 1

BOX – 1

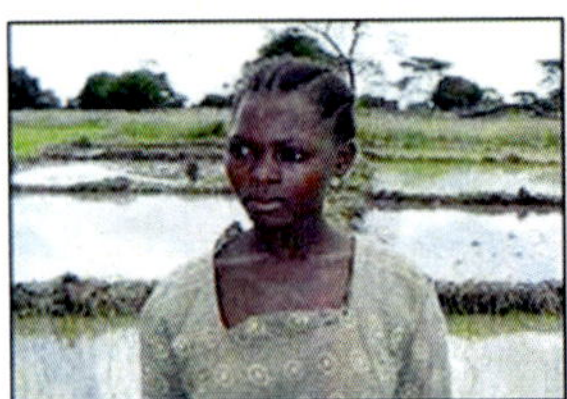

African agriculture is in crisis, and Africa's farmland is losing its fertility at an alarming rate. Farmers grow the same crops in the same fields year. They don't fertilize, and they don't terrace their fields. When the soil wears out, they clear a new piece of land. (wwww.npr.org)

So the crisis of soil fertility loss is the first big problem. Africa's nutrient depleted soils require replenishment. Further, African farmers have to pay two to six times the average world price for chemical fertilizer as given by International Fertilizer Development Corporation.

Such a situation convinced us about the urgent need for an Agro Ecology or Eco Agricultural approach. As given in Box 2 similar sentiment is expressed by the committee World Food Security (CFS)

BOX – 2

Agro Ecology (Eco Agriculture and Small Farmers)(CFS – Committee on World Food Security)

Ref. Ecology / KirtanaChandrasekharan et al at "Friends of Earth"

Agro Ecology that sequestes carbon into soils, making them more fertile, productive and resilent, while also supporting sustainable livelihoods and tackling climate change"

In the bleak scenario, one of the brightest rays of hope is that a dramatic change in our farming system can significantly reduce emissions and put an end to hunger.

"Agroecology – an agenda to focus on ecological farming in the hands of small food producers – is receiving mainstream support".

"There is an alternative ! Small scale farmers – food producers remains the most important and dynamic part of food system. 70-80% of the food consumed in the developing world is consumed by small holder producers and workers. They play a vital role in feeding the most marginalized people – with significantly fewer resources.

Small holders themselves are the most important investors in agriculture, dwarfingany investment in agribusiness. Farmers in low &mild income countries invest more than $ 170 billion/year in their farms – about $ 150/farmer. This is 3 times as much as all other sources of investment combined.Millions of small producers are already practicing agroecology whether they know or not .

The small farm holders can play the most significant role if given an alternative, appropriate tool especially eco-friendly agri bio input started becoming more and more clear as will be explained later. (In this chapter as well as in all other chapters, the name of the products used are based on actual field experiments. They do not, as earlier mentioned in the disclaimer, imply approval to the exclusion of other products that may also be suitable)

OUR EXPERIENCES

ECO AGRICULTURE AND AGRI BIO INPUTS FOR AFRICAN FARMERS

Mauritius – the small African Island is well known for its Sugarcane farms which is the main agricultural activity. Nearly 72000 ha i.e. 85% of cultivated land come under Sugarcane plantation and the local Sugar Industries produce more than 600000 ton / year Sugar of which nearly 17% is exported. During several interactions in early 2000, with the Ministers, Farmers, Co-operatives and Industries it became clear that there is a good awareness and a welcome attitude for eco-friendly farming and inputs. With a local co-operative marketing company, we took up several initiatives of field trials and demonstration.

(a) Treatment of Press mud (locally also called Scum) and other Sugar industry wastes to uniform quantity and highly enriched Bio Composts after Multi Microbial two step treatment.

(b) Using the Bio Composts alongwith a complete package of Bio Fungicide, Bio Pesticides for Vegetable farms as well as Sugar farms.

In Figure 1, the two step treatment of the Press Mud in a typical location is shown. (Details of the two step Multi Microbial process are discussed in a special chapter on Bio Composts later).

Table 1: Enriched Bio Compost from Agro Waste

Tests	Untreated Press Mud	Wonderlife : 1-2-3 Treated Press Mud
pH	8.6	6.8
Nitrogen	1.02%	2.58%
Phosphorous	2.00%	4.70%
Potassium as K	0.43%	1.48%
Sulphur as S	0.10%	0.84%
Moisture	>55%	45 – 50%
Trace Micronutrients	Average	High to Very High
Smell	Pungent	Ammonical
Microbial Counts	1.6×10^3	$>1.0 \times 10^9$

Further, Table 1 lists the analysis of treated Press Mud. There was clearly a remarkable improvement in the quality of the Press Mud which became pathogen free and ready to be applied as a very useful bio input (Ref. 1) In view of the growing interests by small vegetable growing farmers, a complete package alongwith the treated Press Mud was taken up which typically included Multi Microbial Bio Fertilizer for Nitrogen, Phosphate, Solubilizer, Bio Fungicides and Growth Enhancer.

Table 2: Field Evaluation data of Wonderlife : 1-2-3 Enriched Bio Compost and applying package of Bio Fungicide, Growth Enhancer & Bio Fertilizers

Crop	Yield Increase%	Disease Control %	Savings in Agri-Inputs %	Increased Effective Profitability %
Tomato	18	85	30	20
Potato	20	88	35	22
Chili	15	75	30	20
Banana*	25	80	35	25
Sugarcane*	30	85	35	28
Onion	20	80	30	25

Notes:

1. All trials with 20% reduced chemical fertilizers of the recommended dosage
2. Single application of Wonderlife:1-2-3 Enriched Biocompost as basal dose in furrow placement; except where (*) indicated
3. Effective profitability calculated on the basis of reduction of chemical inputs including fertilizers, pesticides and fungicides
4. % figures rounded up for simplification

Table 2 lists the average results of the farm trials. With the support of local scientists in Government departments, the demonstration became very valuable and get the farmers interested.

Bed preparation and applying Bioproducts - Madagaskar

Tomato – Improved product – Lower input cost
Field in Mauritius

This, followed by the involvement, local co-operative marketing unit soon became popular with a large number of farmers, nurseries and green houses. As was visualized1, Mauritius is emerging as a model for other African countries for Eco Agriculture.

UGANDA AND ROCK PHOSPHATE DEPOSITS

In 1992, the Vice President of Uganda visited GSFC Ltd. (Gujarat State Fertilizers Ltd. – Vadodara) and as I was showing him around the various plants and our R & D establishment, suddenly pressed my hand and asked to visit Uganda and the Rock Phosphate deposits at SukuluTororo near the Kenya – Uganda Border and perhaps establish a Phosphatic Fertilizer Complex there!

During my several visits with my team and local scientists at this beautiful site we observed that the igneous deposits of low grade rock phosphate was being considered for a Chemical Fertilizer Complex mainly to produce Triple Calcium Phosphate (TCP) and a project for nearly 120 m US $ was earmarked with World Bank. We took up several studies of the rock phosphate in Indian labs to check the technical viability. Whereas the chemical route to produce TCP did not appear a very attractive proposition, the alternative route of solubilisation of low grade rock phosphate by microbial media was an alternative which looked worth examining.

Samples of Uganda rock phosphate were also taken up for trials for a joint research work by Sambalpur University, Odisha–India who have reported interesting results. The phosphate solubilizing activity of Pseudomonas Fluorescence showed that P_2O_5solubilisation upto 56 (mg%) of P_2O_5 was obtained.

Effect of variables like pH, pulp density, carbon and nitrogen sources and temperature and incubation period were studied and it was concluded that the use of Pseudomonas Fluorescence strain as a phosphate solubilizer is promising, economically attractive and eco-friendly. Our recent studies on using Multi Microbial system has been giving even more promising field results.

Phosphorous is a major nutrient along with Nitrogen and Potash, and the plant growth is affected badly by P deficiency. However, a large portion of phosphate applied to soil as chemical fertilizer is immobilized and becomes unavailable to plants. The application of less expensive natural resource like rock phosphate with Microbial Solubilisation of rock phosphate is thus very worthwhile alternative for Africa.

ZAMBIA AND SOUTH AFRICA

Combining with the teams of local farmers and companies, the teams of scientists from The Science Ashram, Gujarat Life Sciences and Mauritius moved to different farms in Zambia and South Africa where conditions and priorities were different. Three typical examples are :

1. Green House application in Zambia: As shown in the BOX - 3 was as per local publication. A range of bio inputs are typical such a Bio Compost for soil enrichment, a Multi Microbial Bio Fertilizer both in liquid and granular forms and Bio Fungicide and Bio Pesticides were used with promising results.

2. Cotton Farm in Zambia : In a second situation – the local co-operative cotton farming body took just one typical granular product made from Bio Fertilizers, Bio Fungicides and Growth Enhancers.

The result of treated and untreated cotton plants are given in Table below.

Treated plants exhibited vigorous growth in terms branching and height which started to show 3 weeks after the application of the product compared to plants in the untreated plot.

Table 2: Yield Results for Cotton
Cotton Plants treated with WL – G
Experimental Farm of Cotton Development Trust, Magoye, Zambia

Plot	Treatment	Pick 1 (kg)	Pick 2 (kg)	Pick 3 (kg)	Total Pick (kg)	Kg/ha
A	Untreated	60	41.5	9.5	111	2740
B	12kg/ha	91.5	45	11	147.5	3642
C	25kg/ha	89	42.2	10	142	3506

All plots were balanced to the same population of plants. Both treated plots had high yields than the untreated plot as shown in the table above. The highest yielding was plot B which had the recommended application of the product.

BOX 3

Green House in Zambia*

New Generation Agri Biotech Inputs

A range of organic products for Green House. Most of the farmers who tried these products are very satisfied.

Mr. Pradeep Nair, who is the Green House Manager of York Farm has accepted to share his comments:

York Farm is an export oriented farm. We have additional challenges when compared to other farms which are not export oriented. These are :

Controlling pest and diseases efficiently with products of low PHI. We are also limited in the range of products used, as these has to be approved by our customers.

Ensuring that there is no unwanted residues on the exported produce.

Achieve a maximum yield to honour our commitments. This is done mainly by keeping our soil fertility to an optimum level.

I have come across some time back with a range of organic products marketed by MCFI Int. (Z) Ltd. / The Science Ashram & Gujarat Life Sciences – India. These are ECOCERT certified products, consisting of insecticides, Fungicide, Nematicide and fertilizer. They are made of beneficial microbes, plant extracts and other materials of pure organic origin. The product profile suits perfectly with what I was looking for, and I have decided to give them a try.

I must admit that the results obtained from these products have gone well beyond my expectations: I managed to keep and enhance my soil fertility by using the GLS ENRICH, SUPERLIFE, WONDERLIFE G and RELIFE. Also,

Contd.

with the addition of good quality compost, which we made with the help of WONDERLIFE 1-2-3, our sol flora and soil structure are to their optimum levels.

Major insects are now under control with NEEM A LIFE, MITE NO MITE and BEAULIFE. Soil borne diseases are controlled with TRICOLIFE.

"All these products have helped. Now I am somehow relaxed as I know that solutions are available for me to achieve my goals".

Ref : News Zambia, 2007

(3) South Africa – Farmwaste: In a very different situation in some South African farms as shown in the photographs, the major interest of a large farmer was application of multi microbial mixture (in liquid form) to convert the agro wastes from a large wheat farm to high value Bio Compost which can be reused and applied to the farm before the next sowing.

Agrowaste from wheat farm

Multimicrobial Mixture : WL 1-2-3 in Water

Application for Bio Compost from Agrowaste

AFRICA AND INDIA

Chakraborty and Mishra have made a detailed study about the rationale for India – Africa collaboration on food security, given their common challenges for hunger, malnutrition and low productivity. Table - 3 gives idea about prevalence for malnutrition. They concluded: "India plays an important role in augmenting Africa's food output by providing low cost technology solutions, building capacity and providing improved agri inputs and machinery".

Table 3: Prevalence of undernourishment (in millions)

	Number of undernourishment			
Country	1990 - 92	2005 - 07	2014 - 16	Change %
Africa	181.7	213	232.5	28%
India	210.1	233.8	194.6	- 7.4%
World	1010.6	942.3	794.6	- 21.4%

Similar thought was also expressed by JakayaMarisho – President of Republic of Tanzania and Chairman of the African Union at the First India – Africa Summit in April 2008 when he said - "India has the technology and skills, which if made available to Africa, it will help implement the African Green Revolution". However, in the same forum it was concluded that "it will be better to plan a Eco-friendly Revolution by-passing the ill effects of Green Revolution (BOX -4)

BOX - 4 "By passing the ill effects of Green Revolution". The Green Revolution in India has been remarkably successful in increasing production of major crops but also came at the price of various environmental problems: Water logging, salinity, alkalinity, soil erosion, contamination of ground water declining water table etc. Therefore, African policy makers must be mindful of sustainability problems associated with any "Green Revolution". The capital intensive approach to increasing crop production is inherently against small farmers because they cannot afford to buy the expensive inputs. In India, small farmers did not benefit much from new technology and the income gap between small and large farmers in fact widened.Also the climate change is a fundamental threat to food security in India/Africa because most farms are rain-fed. (Ref. :First India – Africa Summit, April 2008)

CONCLUSIONS

It is clear that agricultural development will play the key role for the future of Africa. We have also observed that bringing Eco Agriculture Revolution will be the best way to achieve the goal of poverty and hunger removal in an environmentally sustainable way. Here the large percentage of small scale farms will be the main hope as they play the most important and dynamic part of food system. Empowering the small scale farmers with Eco Agriculture products and technology have been clearly demonstrating the win – win situation. Again these have been demonstrated in different parts of the continent and can become valuable experience for confidently moving ahead.

For the farms, actual field scale demonstrations are the best way to move forward and adopting technology. However, the next big and challenging task is multiplying such efforts and reaching the smallest and poorest farmers.

REFERENCES

1. Mehta M.H. , Key Note Address – First International Agri Business Conference, 25 – 28 September 2008, Mauritius.
2. NEPA (New Partnership for Africa Development), Report on Agriculture in Africa – 2013
3. Panda Ranjit et al; Solubilization of Uganda Low Grade Rock Phosphate by Pseudomonas Fluorescence, J. Science; Vol. 2; 2013.
4. Malancha Chakraborty and Vidisha Mishra, "India – Africa Partnership for Food Security", ORF, Occasional Paper, June 2016
5. Report : First India Africa Summit, April 2008.

"The most significant creative activities of our human culture were made possible through the collaborative work of the left and right cerebral hemispheres. These creative acts, even if by only a few have changed us and the world"

– Carl Sagan

5

The 20 : 20 Model

HIGHER FARM PRODUCTION WITH LOWER INPUT COST

There is a great need to reduce the input costs and at the same time improve the farm productivity in a sustainable manner. A model has been worked out based on a number of field demonstration studies in different parts of India, Africa and Far Eastern countries. A model in which input cost can be reduced by nearly 20 percent and at the same time the crop production can be enhanced upto 20 percent is possible and can be a boon to farmers in most parts of the world. This is the model, which can be easily adopted in an evolutionary way. As suggested by Jonathan Foley (BOX – 1), use of high-tech, precision farming systems, as well as approaches borrows from organic farming, we can boost farm productivity substantially.

BOX – 1

(Ref. National Geographic Magazine, May 2014 – A Five Step Plan to Feed the World by Jonathan Foley)

Unfortunately the debate over how to address the global food challenge has become polarized, pitting conventional agriculture and global commerce against local food systems and organic farms. The arguments can be fierce, and like our politics, we seem to be getting more divided rather than finding common ground. Those who favour conventional agriculture talk about how modern mechanization, irrigation, fertilizers, and improved genetics can increase yields to help meet demand. And they are right. Meanwhile proponents of local and organic farms counter that the world's small farmers could increase yields plenty – and help themselves out of poverty – by adopting techniques that improve fertility without synthetic fertilizers and pesticides. They are right too.

But it need not be an either or proposition. Both approaches offer badly needed solutions; neither one alone gets us there. We would be wise to explore all of the good ideas, whether form organic and local farms or high-tech and conventional farms, and blend the best of both.

Among 8 million goals, eradication of poverty and hunger and environmental sustainability are perhaps the most prominent and complementary. The Green Revolution changed India from (& many other countries) being food deficit to food sufficient / surplus state. Time is now for a new and Sustainable Revolution.

If we look at the challenge that agriculture sector has to face during the next few decades, it is clearly to double the food production by 2025 and triple it by 2050 on less per capita land, with less water and under environmentally challenging conditions.

Let us look at the current scenario and some major developments.

- Indian farmers spend nearly Rs.2 Lakh crores (1 Lakh = 100000 & 1 Cr = 10 Million) on inputs like seeds, fertilizers, pesticides.
- Nearly Rs.50,000 crores goes as fertilizer (chemical) subsidy.
- Excessive use of chemical fertilizers in India hasreduced soil fertility and food production (Green Peace &Vishwa Bharti report – 2009). Fertilizer use efficiency is drastically declining. (It has already come down from 19 Kg food grains per kg fertilizer in mid 60s to 12 kg/kg in 2007).
- Our soils are hungrier for organic carbon. The carbon transfers and transformations at the soil or even the cell level can delineate the other nutrients such as N, P, S etc that become available to plants and microorganisms.

There is now a better understanding of Multi Microbial Consortia Technology which can substantially improve crop productivity and at the same time reduce the agrochemical input costs. Further, new method of preparing high quality Bio compost from agro-wastes, FYM, etc can greatly improve the availability and quality of nutrients of assured standard and much needed Carbon to the soil. Another startling development has been in the area of pesticides. In January 2009, the European Council passed the Council Directive 91-414 banning nearly 45% of the prevailing chemical pesticides. This will have great impact on the overall world agricultural scenario. As a matter of fact, in 1993, nearly 740 active ingredients were available in European market whereas on August, 2008 nearly 394 of those were not allowed and 113 more are under review (FCI – December, 2008). Two major classes of fungicides at risk are triazoles and dithiocarbonate. Triazoles are main weapon against Septoria – a major fungal disease in cereal. The Swedish Chemicals Agency, a Govt. organization responsible for a non-toxic environment was one of the first EU member to interpret the temporary criteria for endocrine disruption and published a 'Black List' of some prominently prevailing chemical pesticides.

All these developments point out to :

- An urgent need for eco-friendly and effective Bio inputs including Bio Pesticides.
- Need for their availability at low prices and easy supply.

Fortunately, several new/alternative developments offer much hope.

Range of Bio Inputs

Broadly Bio Inputs range can be classified under three main categories :

1. Nutrient Management: Biofertilizers, Enriched Biocompost, Multi-Microbial Combination of Bio Fertilizers and Bio Fungicides in the soils.
2. Bio Pesticides: Multi Microbial Bio Pesticides, Botanicals, Pheromones etc.
3. Growth Enhancers: Amino acids, micro nutrients, seaweed extracts, growth promoters and hormones, etc.

In Annexure – 1: The typical range of products is given.

It is possible to enhance the farm productivity and at the same time lower the input costs through the use of agri biotech products. The range of multi-microbial consortia products play the key role in these combinations.

Field Demonstration and Triangular Arrangement

A triangular arrangement involving an Agricultural University, a private biotech company and an NGO was thought to be the best way for taking up demonstrations of the new generation agri biotech products for practical field applications. A program was taken up where Navsari Agricultural University (South Gujarat), (which is a leading Agri University with an experimental farm and specialized group of scientists), alongwith Gujarat Life Sciences (Pvt.) Ltd. (GLS) – a new generation agri and enviro biotech company and the NGO – The Science Ashram with a network of farmers and links with industries and universities was made for such a program.

Annexure –1

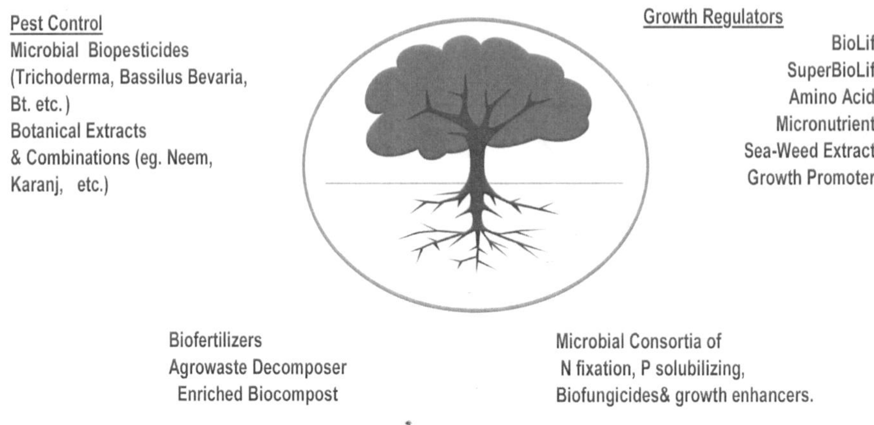

*Specialized Formulations

Annexure – 2

Various products were taken up for field demonstrations, applications are listed (Annexure – 2). The typical products involve – enriched bio compost, bio pesticides, bio fertilizers, etc. The list is typical and can be taken up as combination from different sources of suppliers also.

Typical productivity Enhancement Data – Multi Microbial Products, Bio-Fertilizers and Biocomposts.

(Application of Bio Products – Superlife, Wonderlife, Bio Compost)

Table 1:

Crop	Location	Result
Banana	Agricultural University- Navsari, Trichy-Tamil Nadu	Above 25% more production with above 20% reduction in urea dose.
Rice	Bangkok – Thailand, Vapi-Gujarat	Increase in production from 14 to 19%.
Maize	Gauteng – South Africa	25% more production, 25% less Urea.
Castor	Kutch – Gujarat	Increase in yield by 17-22%.
Papaya	North Gujarat	Up to 21% increase in yield achieved.

Annexure – 3 : Typical Field Demonstration Data

In Table 1, a typical productivity enhancement data for different crops and different data are listed. Incidentally there may be variation for different crops and different locations, but clearly it is established that there is :

(a) Enhancement in productivity, and (b) Reduction in input costs of agro-chemicals. Similarly, in Table 2, typical data on disease control for various crops in different locations by use of Bio Pesticides and Bio Fungicides show excellent results. It is thus clear that a comprehensive model can be made available to the farmers, where substantial reduction in input costs and at the same time improvement in the productivity can be achieved. That this can be done through eco-friendly inputs is another big advantage for sustainable agricultural practices.

It may be noted that in explaining this model, the data for some typical bio products are given as indicative. Right combinations from any standard suppliers can be used. Farmers can also develop their own Bio Composts from their farm wastes, FYM etc by using standard microbial cultures.

Typical Data Improvement

Disease Control – Use of Biopesticides, Biofungicides, (Tricholife (Bio fungicide), Superlife, Wonderlife-G, Neem-A-Life, Mite-no-Mite (Bio pesticide).

Table 2:

Crop	Location	Result
Brinjal	Saurashtra / Trichy – Tamil Nadu	Use of Mite-No-Mite has given control over mites and upto 15 days there was no need to do any other spray for mites.
Cotton	Central Gujarat Alwar - Rajashtan	80-85% control in various pests using Neem-A-Life
Groundnut	Saurashtra	90% control in fungal diseases with Tricholife + Superlife
Roses	Mauritius – Central Trichy – Tamil Nadu	Use of Mite-No-Mite 90% control on Mite problem. Excellent Growth.
Potato	Central Gujarat	By using Wonderlife-G in Soil, disease control upto 94% and upto 22% increase in yield is achieved.

THE 20 : 20 MODEL AND IMPLEMENTATION PROGRAM.

In the recent past, the 20:20 model is emerging as perhaps the most practical and popular model with the farmers. It is realized that though the environmental problem and the high costs of chemical farming are noticed by all, they cannot be wished away or stopped overnight. On the other hand, farmers who suddenly tried to switch over to organic farming and faced disappointment and problems which did lot of disservice. Between the two, 20:20 model has shown the practical and positive way and easy adoption. It is also seen that after the first year of 20:20, the bioorganic inputs use can easily be doubled and even tripled in the next two years by many progressive farmers. Such farmers are in different agro-climatic conditions and with different crops.

BOX – 2

Nowadays, many countries, including developing countries are all experiencing difficult transition from conventional agriculture to environmental friendly agriculture. To balance high inputs, outputs and maintain ecosystem health become a huge challenge for each nation. This 20:20 model with higher production and lower input costs will surely bring a bright future for the agriculture in the world.

Prof. Li Ji, Beijing Agricultural University.

It was, therefore, heartening when at an international conference in Beijing, a group of practicing farmers and scientists of China told me about their happy experience with such a model and their intention to spread it in a much bigger way.

CONCLUSIONS

(i) There is a dire need of improving farm productivity and equally the need to reduce the input costs.

(ii) The 20:20 Model can be taken up to substantially help the farmers in meeting such a goal.

Applicability of such a model is even more urgent under the present condition of water shortage and food crisis.

"A Small needle can do what a sword cannot"

– Rahim

6

Bio Fertilizers – Industrial Perspective

To many Bio Fertilizer is still an unclear term. Often they are identified as urban and rural organic wastes, plant extracts, organic components etc. However, Bio Fertilizer is mostly commonly referred to the use of soil micro organisms to increase the uptake and availability of plant nutrients.

Bio Fertilizer have gained a good deal of momentum and the industry after a slow start is now gaining more and more importance globally. The major reasons for this are that they:

1. Improves the soil structure.
2. Add nutrients to the soil / make the nutrients available to the crop
3. Secret certain growth promoting substances
4. Under certain conditions exhibit antifungal activities and there by protect the plants from pathogenic fungus
5. Enhances seed germination
6. Increase the soil fertility
7. Increase the agricultural productivity
8. The nitrogen fixed or phosphate solubilized will be beneficial to the subsequent crops also.

The major, popularly used and commercially produced Bio Fertilizers are Nitrogen fixing bacterial fertilizers (Azotobacter Culture, Azospirillum Culture, Rhizobium Culture) and Phosphate solubilizing bacterial fertilizers (Pseudomonas Culture). However, currently their applications for other useful elements like Potash, Sulphur, Zincetc are also increasing.

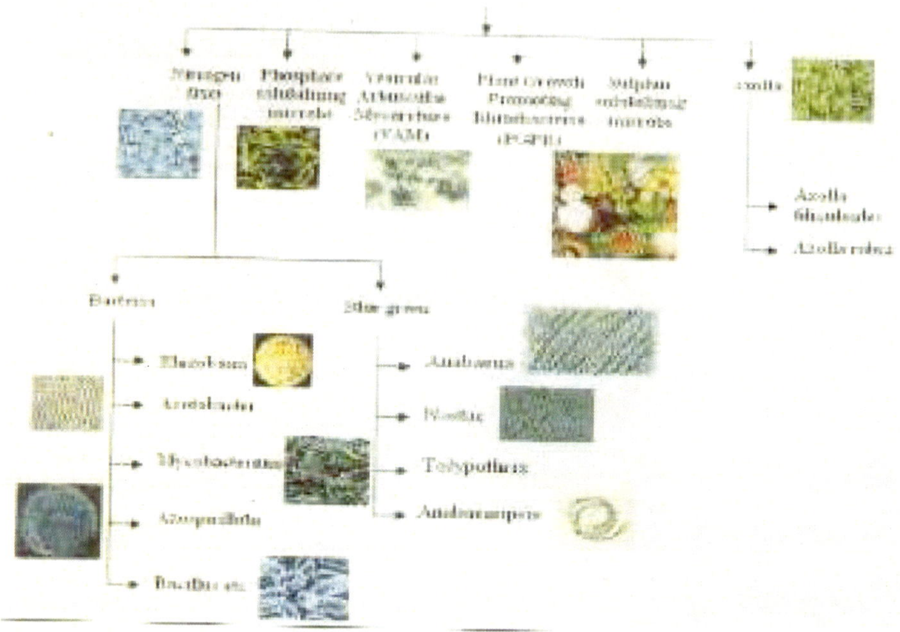

Fig. 1: Classification of Bio Fertilizers

TYPES OF BIO FERTILIZER

In Fig. 1, based on their nature and functions,the broadclassification is given. Of these, the mode of action of the major Bio Fertilizers are as follows :

1. **Azotobacter Culture:** They are the non-symbiotic (free living) nitrogen fixers. They are isolated from the rhizospheric soil of any cereal crop. The nitrogenase enzyme present in the bacteria is responsible for nitrogen fixation.

2. **Azospirillum Culture:** They are the associative nitrogen fixers. They characteristically remain in contact with the root surface. They are also isolated from the rhizospheric soil of any cereal crop. They also possess nitrogenase enzyme for nitrogen fixation.

3. **Rhizobium Culture:** They are the symbiotic nitrogen fixers. They require specific host for nodulation. They enter the root system of the specific legume plant and form nodule. They multiply there and provide nitrogen to the plant while in turn they get carbohydrates from the plant for their multiplication. They also possess nitrogenase enzyme for nitrogen fixation.

4. **Pseudomonas Culture:** Certain specific strains Pseudomonas secret certain organic acids which dissolve the bound unavailable phosphate (Tri-Calcium Phosphate) of the soil and make the phosphorous available to the plant.

Another important bio input for sustainable farming is Mycorrhizae (Fig. 2 & Box) Arbuscules Mycorrhizal Fungi (AM) and with a mutuality relationship with roots of most plant species. The importance of these fungi to agricultural

Mycorrhiza with Intracellular Haphae

Mycorrhiza with Arbuscules formation

Fig. 2: Pictures of Mycorrhizas

and forestry residue in their role in plant growth and nutrition. Due inoculation of such fungi with Rhizobium and other bacteria on plant enhanced the growth and other benefits viz. resistance to disease and tolerance to adverse soil and climatic conditions.

BOX — 1

Mycorrhiza – An important Bio input

Mycorrhizas are indigenous to soil and plant rhizosphere and are potential tool for sustainable agriculture. They enhance the growth of the plant and control plant pathogens.

Mycorrhizas is emerging as an important input as it helps promotion of plant health, soil health and soil fertility.

Mycorrhizas are obligate and saprophytic in nature which are totally biotroph and hence require a living host for its survival. Mycorrhizas by associating symbiotically with root of the plants helps in the greater absorption of phosphorous, water, other micro and essential micro elements and making them available to the plants in usable form. They also impart resistance to the plants against drought and soil borne fungal pathogens and nematodes.

MULTI-MICROBIAL FORMULATIONS

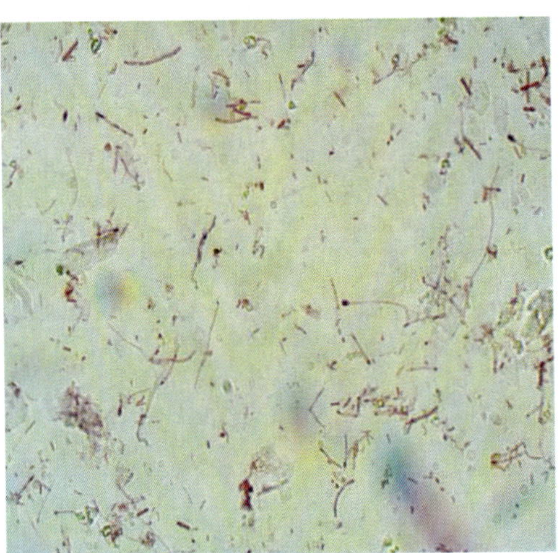

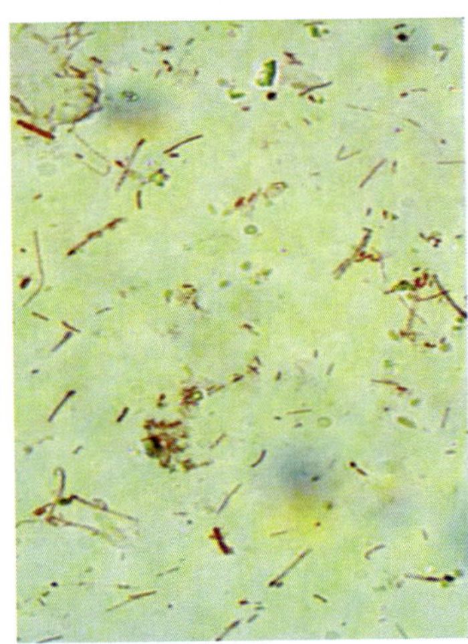

Fig. 3: Monoculture and Multi Microbial formulation

Bio Fertilizers based on Multi Microbial Consortia Technology is gaining ground as more advantageous and superior technology. Bio Fertilizer based on simple microbe and inert material (solid or liquid) as a carrier perform a simple function like Nitrogen fixation, P Solubilization etc.

Multi Microbial Consortia offer several advantages like :

1. They are economical to farmers
2. Easy to use as a single application
3. Help plant to assimilate more than one macro or micronutrient as compared to monoculture microbes
4. With the help of Multi Microbial Technology, it is easy to make products which help in disease and nutrient management of crop.
5. Shelf life of Multi Microbial Technology based is more as compared to monoculture bio fertilizer due to less contamination. As Multi Microbial based Bio Fertilizers have group of microbes which are compatible with each other. the metabolites produced by one microbe are used for the growth of another microbes.
6. With the help of Multi Microbial Technology, it is easy to make broad spectrum formulation / products.

SCALE UP AND MASS PRODUCTION

Over the past few decades, the Bio Fertilizer production has moved from small lab scale unit to industrial scale production. Figure 4 gives the stepwise approach.

Isolation of Microbes from roots or soils

Laboratory screening of microbes for plant growth

Green House screening of microbes to promote growth in potted soil

Field screening of most effective microbes cropped soil (Crop variety and different soil types examined)

Refinement of Inoculums

Environmental impact test and substantiation of microbes

Production

Fig. 4: Bio Fertilizer Production Steps

In selection of strain for mass production, the two important criteria to be considered are; the ability to form effective nodules with the hosts for which the culture is recommended and the ability to possess this trait under a wide range of field conditions. The ability of the new strain to compete out the native strains in soil is perhaps a foremost consideration in selecting a strain. Other important criteria are prompt nodulation over a range of root temperatures, greater survival in carriers and seed, and tolerance to pesticides, chemical fertilizers and other agrochemicals. The use of 'multi-strain' inoculants may be considered if effective strains for a limited number of hos species have been identified by field tests. Also it is important that the mixed strains when added to carrier (such as lignite, peat etc), must posses equal growth rates.

The primary inoculum for seeding in fermenters should attain the maximum possible population of microorganism within a short time (usually 3 to 9 days). The size and design of fermenters and factors like aeration, initial inoculum level, bacterial strain, temperature and incubation time etc are the important factors affecting output of cells.

Research and development program concerning the production of biomass have greatly influenced technological progress in the fermentation industry. Overall process economics are affected by both physiological and engineering factors.

1000 liter fermenter

200 liter fermenter

Fig. 5: Fermenters

The main physiological factor that affects process economics is the yield coefficient, defined as the weight of dry biomass produced per unit weight of nutrient utilized, while the main engineering factors that affect process economics is the fermenter productivity, defined as the weight of dry biomass produced per unit fermenter volume per unit time.

Impeller agitated, spurge, baffled, cylindrical fermenters have dominated the fermentation industry since the industrial scale antibiotic production by submerged culture technique was introduced in mid 1940s. Till 1970 in the design of industrial fermenters multiple impellers which led to an increase in the aspect ratios of impeller agitated fermenters were dominant. More recently, several alternative fermenter designs of improved economics are claimed. These designs can be classified into three groups namely – mechanically mixed systems, hydrodynamic systems and pneumatic systems. It may however be emphasized, that looking at the today's development state of Bio Fertilizers, the conventional impeller agitated fermenter will continue to be used because of its overall flexibility.

The Fertilizer Control Order of Govt. Of India has published the specification for Bio Fertilizers in 1985 which has been revised / amended from time to time.

BIO FERTILIZER MARKETS : NATIONAL AND GLOBAL TRENDS

Marketing and Global Scenario

When we visualized the great potential for agriculture in early 1980s at the R& D and Agri Service facilities of giant chemical fertilizer company GSFC Ltd., there were apprehension and questions about its relevance and future. Chemical fertilizers were in big demand and it was 'sellers' market. The idea of producing Bio Fertilizers seemed utopian and impractical to many. Even, greater was the apprehensions. Educating the Government Bodies on one hand and farmers on the other hand looked a daunting task.

It had been recognized since last several decades that scaling economic and energy costs and environmental considerations about chemical fertilizers had drawn attention towards Bio Fertilizers. GSFC Ltd. – a large joint venture corporation of the Govt. of Gujarat which was one of the largest producers of a whole range of chemical fertilizers which were in much demand. So when we decided to start work on Bio Fertilizers there was much apprehension. First, it took time to explain to everyone including the Board of Management as to "why" do Bio Fertilizers when our chemicals were in a great demand and growing. It even took a some efforts to fund R & D work on Bio Fertilizers and Bio Pesticides. However, the usefulness was soon shown and first commercial scale unit was soon established and marketing took up. This is a typical story of the beginning of commercialization of Bio Fertilizer units which I have later compared and found many parallels in other organizations in India and Asian countries.

In the meanwhile, Government of India's Ministry of Fertilizers and Chemicals formed a National Committee on Bio Fertilizers in late 1990s under my Chairmanship to draw out a future plan for actions. We took up exercises on aspects like developments, standardization and specifications the future field applications and marketing and a road map was prepared.

Working with reputed Market Consultants, it was also concluded that nearly 56% of farmers became aware about Bio Fertilizers through Farm Information Centers and about 17% through advertisements and the rest through other sources including ear say. Further, the users of Bio Fertilizers grew in two years time as given in Table below :

Table 1: Use Of Bio Fertilizers By The Farmers – Early Years (First 4 years of introduction after 1994)

Category of the farmer	Not used Bio Fertilizers (%)	Used Bio Fertilizers (%)	Stopped using Bio Fertilizers (%)
Marginal	56	22	22
Small	17	62	21
Medium	12	71	17
Big	10	66	24

Another interesting development happened when during one year it did not rain enough and the Government of Gujarat decided to distribute a large quantity of Bio Fertilizers to the farmers in rain fed areas as a major of relief . Suddenly, the demand for Bio Fertilizer packets increased in a big way that year and our production teams had to work over time ! This also gave a major boost to the awareness drive. Today, many states in India, promote Bio Fertilized application through respective Government support.

GLOBAL TRENDS

In the last 2 decades, the Bio Fertilizer industries globally have shown promising trends. Based on recent market surveys the following observations are made :

A) As the biofertilizers market is growing due to increased awareness among consumers regarding chemicals and a growing inclination towards consuming products that are free of chemical and synthetic ingredients. An increase in government funding, support towards the use of biofertilizers, and the growth of the organic food market are some of the major factors driving the global biofertilizers market. Moreover, farmers themselves are limiting or even discontinuing the use of chemical fertilizers as their long term ill-effects have come to the fore. Heightened awareness among people regarding the detrimental effects of chemical fertilizers on the environment is also expected to benefit the market.

B) According to the report, the global biofertilizers market will expand at a strong 13.0% CAGR from 2013 to 2019. The market's valuation was US$440.0 bn in 2012 and is anticipated to reach US$1,028.7 bn by 2019 owing to a growing awareness among people regarding the environmental hazards associated with the use of chemical fertilizers.

C) Biofertilizers have not yet been used consistently across the world, which can impede the market's growth. In terms of opportunities, biofertilizers have a good scope of development in emerging countries such as China and India where agriculture is still a major part of the country's economy.

D) On the basis of application, the market is broadly categorized into seed treatment and soil treatment. The report states that in 2012, the seed treatment segment was the largest and was worth US$316.5 mn. Its growth can be attributed to the growing restrictions by governments on the use of chemical-based pesticides, compelling a large number of farmers to adopt eco-friendly treatment options for seeds. For the same reasons, the soil treatment segment is anticipated to witness great demand in the forthcoming years.

E) Due to the growing demand for organic food, North America held the largest share in the global biofertilizers market, followed by Rest of theWorld. However, Asia Pacific is anticipated to witness the highest growth rate, thanks to an increased awareness among people regarding the ill-effects of chemical fertilizers. The Asia Pacific market is expected to expand at a CAGR of 13.3% in the period from 2013 to 2019.

CONCLUSIONS

Globally the demand for Bio Fertilizers will grow remarkably. The development of formulations based on Multi Microbial Consortia will play an important role. This combined with a total package of agri bio inputs will be the future trend.

REFERENCES

1. Mehta M.H. 1997. Keynote - Bio Fertilizer – Industrial Perspective ; TCDC International Workshop on Bio Fertilizer at New Delhi, October 16-17.
2. Manisha K. et al, 2015. Chemical Weekly.
3. Hari Muralidharan et al ; 2010. Bio Fertilizer Booklet; Amon M. Chatterjee Research Centre, Chennai.

4. Pushpa Nair et al 2011. (Gujarat Life Sciences); GLS Report 2011.
5. Mehta R.D. Patel RM and Mehta MH, 1983. Bio Fertilizer – ISTE Winter School, M.S. Uni. Baroda
6. Transparency Market Research 2014. New York, May 7, 2014 (www. transparancymarketresearch. com/biofertilizer-market)

"We see a strong and dominant role of Biopesticide companies for crop protection and production over time".

— Pam Marrone
CEO – Marrone Bio Innnovations

7

Bio Pesticide Industries Global Opportunities and Challenges

At an Agri Summit, I had a very difficult task to chair a session on "Sustainable Future", with two well-known but extremely opposite personalities. On my left was a globally well-known environment activist talking about all the evils of chemical fertilizers and chemical pesticides. On my right was the President of Pesticide Association with equally strong views about how the chemical pesticides have greatly helped in feeding the hungry world. My task was further made difficult by a large lobby of green activists in the audience and also a number of seasoned industrialists with opposite views. Such a scenario is becoming increasingly common.

In January 2009, the European Council Directive 91-414 was passed banning a total of more than 40% chemical pesticides (David Frabotta, FCI – March, 2009). Earlier in 1993, there wre 740 active ingredients available in European markets. By August 2008, however, 394 of these were no longer allowed and 113 more were still under review. It was predicted that in next 5–10 years, 85% of these could disappear from European markets. The Swedish chemical agency shortlisted nearly 21 chemical under criteria of 'Endocrine' (the Endocrine criteria are defined as any substances that trigger class – 3 carcinogenic effect and class - 3 reproductive effect). These are new herbicides like: loxynil, Gluctosinate, etc. The two major class of fungicides are described as Triazoles and Dithiocarbonate. Triazoles are major weapon against septoria – a major fungal diseases in cereals.

BEYOND GREEN REVOLUTION

In spite of increase in the use of agrochemicals, the agricultural production and productivity has remained stagnant beyond 1990 due to poor soil conditions and pests and diseases causing over Rs.300 million in crop losses per annum. On the other hand the contamination by residue of chemical pesticides in food, milk, river and water sources have been alarming. The world wide awareness of harmful effects of organo chloride and organo phosphorus pesticide is clearly making impact on the chemical pesticide industries.

PESTICIDE 'RICH' FOOD

In 1962, Rachael Carson published the book 'Silent Spring" bringing out the ill effect of chemical pesticides on environment, public health and all the life system. Since then, we have been regularly getting stories about such problems. In Table 1, (Pesticide Rich Food published in 'Down to Earth February 2011', the results of study by Department of Agriculture and IARI for different food items indicate the kind of health impact of various chemical pesticides popularly being in much use in India:

We find here that not only vegetables and fruits, but poultry, milk and butter carry alarmingly high pesticide residues. The task of finding safer alternatives, which are eco-friendly, cost effective and productive becomes extremely important and challenging.

BIO PESTICIDES

Biopesticides have drawn much attention in the last couple of decades as big alternative hope. The concept of biopesticide has often been described in the following manner:

Biopesticide or natural pesticides are reduced-risk products derived or developed from biological or naturally derived chemistry. Biopesticides offer value to users by providing a combination of both effective performance and product safety. Biopesticides perform efficaciously while providing the flexibility of minimum application restrictions, superior residue resistance management potential, and human environmental safety benefits.

Some of the unique features and benefits of biopesticidesinclude:

- The ability to provide alternative modes of action to traditional products which makes them a critical component in most IPM programs
- Registration in less time than conventional chemical products because biopesticides exhibit minimal environmental and humans;
- The ability to extend the life of conventional chemicals by providing resistance management benefit programs ;
- Exemption from tolerances, such as reduced pre-harvest restrictions and application in environment which permits biopesticides that have no Maximum Residue Levels (MRLs) to be used on crops in urban settings;

Table 1: Pesticide Rich Food

Food Item	Pesticide Found	Health Impact	MRL – ppm (Max. Res. Level)	HRM – ppm (Highest Res. Level)	Place
Cauliflower	Chloropyriphos	Nerve system	0.01	1.171	Amritsar
Poultry	Endosulfan-T (Rest. Use)	Neutrotoxin	-	0.65	East Nynvau
Milk-Amul	Chloropyriphos	Nerve system	0.01	2.36	Ahmedabad
Butter	Heptachlor (banned)	Nerve system, Liver	0.15	0.51	Faizabad
Cardamom	Qunilphos	Endocrine disruption	0.01	0.15 to 0.25	Idduki
Brinjal	Heptachlor (banned)	Nerve system, Liver	0.05	0.48	Allahabad
Apple	Chlordane (banned)	Nerve system, Liver, Kidney, Eyes, Lungs	0.1	0.93	Gorakhpur
Tomato	DDT (Rest. Use)	Carcinogen, Reproductive system	0.001	0.18	Allahabad

Ref : Down to Earth, February 2011

- a high degree of worker safety and the shortest recently intervals allowed by law;
- value added benefits, such as improved plant health, yields and quality and an increase in beneficial and organic cultivation programs

In the Table – 2, a comparison of biological v/c chemical pesticide gives a good idea about the greater need for biopesticides and biocontrol agents.

Biopesticides can be used in almost any crop production program because they offer unique modes of impact on the environment and human health. They are especially suited for use in :

- Rotation with chemicals in traditional programs to manage for pesticide resistance.
- Certified organic production systems
- Grower programs where pesticide residue management is important for harvest management
- Crops with intensive labor demands to gain maximum flexibility in managing work crews.

Consumption of bio agents like entophages, microbial and botanical pesticides, pheromones is merely compared to chemical pesticide, but rising.

Table 2: Comparison – Biological v/s Chemical

Category	Biological control	Chemical control
Environ. Pollution – danger to man, wildlife, *etc*.	None	Considerable
Upsets in natural balance, ecological disruption	None	Common
Permanency of control and economics	Permanent & therefore economical	Temporary. Repeated application required
	Initial cost low	Costly
Development of resistance	Extremely rare	Common
Income on pests becomes effective	Takes time, but may rebound	Speedy result
Range of Action	Specific or narrow	Broad

BANNING AND BLACK LISTING OF CHEMICAL PESTICIDES

In January 2009, the European council passed the directive 91-414, banning a nearly 45% substances as Pesticides. Endocrine criteria – designed as any substance that trigger class 3 carcinogenic effect.

Table 3: Black Listed

Active Ingredient	Year of Expiration
HERBICIDES	
Armitrole	2011
Ioxynil	2014
Glulufosinate	2017
Linuron	2013
Molinate	2014
Active Ingredient	Year of Expiration
FUNGICIDES	
Carbendazim	2009
Dinocap	2009
Flumioxazin	2012
Iprodion	2013
Mancozeb	2015
Maneb	2015
Active Ingredient	Year of Expiration
INSECTICIDES	
Bifenthrin	2018
Lufenuron	2018
Thiacloprid	2014

Ref : Swedish Chemical Agency - 2009

Recently Govt. of India has published the list of several pesticides (Table – 4) which will be phased out.

Table 4: Chemical Pesticides to be phased out - India

Name of chemical pesticides	Name of chemical pesticides
Benomyl	Metyhl Parathion
Carboryl	Sodium Cyanide
Diazinon	Thiometon
Fenarimol	Tridemorph
Fenthion	Triazophos
Linuron	Alachlor
Methoxy Ethyl Mercury Chloride	Trichlorfon
Dichlorvus	Phorate

Ref : Govt. of India – Ministry of Agriculture Notification, January 2017

Such trends are increasing around the world and the urgency of safe, eco-friendly and effective Biopesticide is being felt across the globe.

COMMERCIAL BIOPESTICIDES

Products derived from biological sources – biopesticides are emerging as new and desirable crop protection strategy. A detailed list of more than 190 Biopesticides by ingredient has been given by US Environ Protection Agency EPA and is available at www.epq.gov/pesticides/biopesticides /productlistindex. htm.

Microbial bacterial cultures like Agrobacterium radiobacter, Bacillus subtilis, Burkholderiacepacia, Pseudomonas fluorescence, etc ; fungi are Trichoderma spp., Pythium oligandrum, Glicocladiumvirens, Fusarium oxysporums, etc.; suchas nuclear polyhedrosis viruses, cytoplasmic polyhedrosis viruses, and granulosis viruses are some of the important biocontrol agents that are available under various trade names.

The most promising botanicals for the present and future use belong to families such as meliaceae, rutaceae, asteraceae, malvaceae, labiatae, and canellaceae (Jacobsan 1989). Plants most commonly used as insecticides are pyrethrums from Chrysanthemum cinerariaefolium (compositae), rotenoids from the genera Derris, Lonchocarpus, Tephrosia, and Mundulea (leguminosae); nicotinoids from Nicotiana species (solanacae), triterpenoids from Azadirachtaindica (meliacea); and alkaloids from sabadilla. All of them, particularly neem, have been found effective against several pests, which include rice cutworm, diamond backmoth, rice brown plant hopper, rice green leafhopper, tobacco caterpillar, and several species of aphids and mite.

The special global interest in neem has been playing a great role in several industrial units manufacturing and formulating these products. Additionally, Govt. of India's decision to promote neem coated Urea has made a significant impact. A key consideration is the stabilization to improve shelf life and field persistence. The product has to be effective under all climatic conditions.

Karanj (PongamiraPinnata) is another important botanical which has insecticidal and bacterial properties and is effective against a wide range of pests and also used as disinfectant for domestic applications.

Working with some rose farmers in Tamilnadu, we heard complaints about problems of mite which was causing a major damage inspite of use of heavy doses of costly chemical pesticides in the market. After several trials, a combination of neem and Karanj successfully solved the problem as shown in Table – 5. What was attractive was that the lower cost and eco-friendly sprays.

Table 5: Mite-No-Mite (a Botanical Bio Pesticide) for roses : Tamilnadu

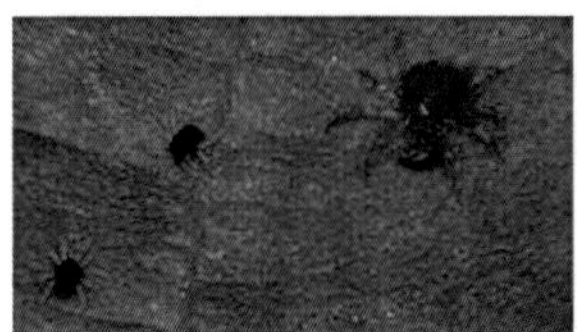

Mite-No-Mite Application on Roses

Sr.	Pre-count of mites per shoot/Plant	Days after 1st Spray		Days After 2nd Spray	
		3	7	3	7
1.	8.30	4.55	3.66	2.45	1.19
2.	8.05	5.42	3.42	2.00	1.08
3.	7.82	4.90	3.00	2.06	0.81
4.	8.98	3.50	2.88	2.21	1.44
5.	7.90	4.25	2.56	1.87	0.24
6.	10.20	5.13	4.33	2.66	2.12
7.	9.00	5.60	3.75	1.73	0.65
8.	8.66	4.41	3.88	2.90	1.20

There are a large number of botanical biopesticides and development and standardization of commercial formulations is a major challenge.

Applications and Demonstrations

Central Gujarat farmers have been heavy users of Chemical Pesticides for cotton and vegetables.

When the young volunteers of The Science Ashram alongwith University researchers started interacting with them in late 1990s – the harmful effects on overuse of chemical pesticides, they listened carefully. Unsure of the 'new inputs' they needed demonstrations and field results.

Table 6: Pigeonpea Recommendation- NAU – GLS – TSA Experiment

Treatment for	Chemical Treatment	Organic Alternative
Macronutrients	Urea, DAP and others	Enriched BiocompostWonderlife-G – GranulesSuperlife through drenching
Growth Regulation	Chemically derived Hormones & regulators	Biolife
Fungal and control	Chemical fungicides	Superlife and Tricholife
Micronutrients	Chemically derived micronutrient	Superlife and specific microbial inoculants depending on specific micronutrient requirement
Heliothis and Spodoptera pests; Aphids and Thrips and sucking pests; nematode control	Chemical pesticides	Pheromone trapsNPV Neem-based formulations Beauveria bassiana Bacillus thuringenesis Bionematicide

Table 7: Overall results

Organic Cotton & Pigeonpea Cultivation

- Cost of inputs decreased from Rs. 3,000 on average to less than Rs. 1,500 per acre.
- Number of sprays reduced from an average of 8-13 to 2-5
- Average cotton production found between 400 - 450 kg/acre
- Population of natural pest predators like lady bug beetle, chrysoperla, trichogramma etc. found in abundance
- No damage to the crops, whatsoever, which may be generally caused due to excessive chemical sprays

Through a triangular arrangement between Navsari Agricultural University (NAU), The Science Ashram (TSA) and Gujarat Life Sciences (GLS) application packages were made for organic Pigeon Pea as given in Table – 6. The general results are listed in Table – 7. Such collective efforts and demonstrations greatly help in building the confidence of scientists and farmers.

Chandra and Greep after extensive studies, have recommended Bio Pesticides and I.P.M products of choice for some important crops as given in Table – 8.

Table 8: Bio Control agents and Bio Pesticides and IPM products for various crops

Crop	Pest/Disease	Biopesticides & Ipm Products Of Choice
Cotton	Bollworms	Traps, Lures, B.T, NPV, Trichogramma
Paddy	Yellow stem borer, Leaf folder	Traps, Lures, B.T, Trichogramma
Mustard	Aphids	Neem 300 & 1500 PPM, Chrysoperia
Brinjal, Bhindi	Fruit Borers, Mites	Traps Lures, B.T, Trichogramma, Neem 300 & 1500 PPM, Verticillium
Citrus	Caterpillars, Whitefly	B.T, Neem 300 & 1500 PPM

They have also listed some major pest and diseases and the Bio control agents and Biopesticides as given in Table – 9

Table 9: Bio Control agents and Bio Pesticides and IPM products for various pests and diseases

Pests/Disease	Biological Inputs
Bollworms/cutworms/Borers	Bacillus Thuringiensis (BT), Beauveria, Trichogramma, Nuclear Polyhedrosis virus (NPV/GV), Chrysoperia, Traps & Lures
WhiteflyTraps	Neem 300 & 1500 PPM & above, Sticky Beauveria, Chrysoperia
Aphids & JassidsTraps	Neem 300 & 1500 PPM & above, Sticky Beauveria, Chrysoperia
Thrips	Neem 300 & 1500 PPM & above, Verticillium, Sticky traps.
MitesVerticillium, Beauveria	Neem 300 & 1500 PPM & above, Verticillium, Sticky traps.
Wilts, Root rots, Leaf spots	Trichoderma, Pseudomonas
Blights, Anthracnose, Mildews	
Bacterial Wilts/Blights	Pseudomonas
Nematodes	Trichoderma, Bionematicide

The above representative examples are typical and world over a great deal of field trials and large scale applications are increasing the confidence and popularity of biopesticides.

BOX -1

More Companies To Be Assessed In Second Sustainable Cotton Ranking

Pesticide Action Network (PAN) UK, Solidaridad and WWF have released the list of companies that will be assessed in the new round of their Sustainable Cotton Ranking to be published in October 2017. The second edition of the ranking will include major companies from all continents, including from countries such as China and Brazil, and online companies such as Zalando and Amazon. As in 2016, the ranking will score companies on their policy, traceability and actual uptake of sustainable cotton

........**COTTON GLOBALLY**

Cotton is grown in around 80 countries worldwide and is a key raw material for the textile industry, accounting for around 32% of all fibres used. Sustainability issues include the widespread use of pesticides, with 6.2% of global pesticide sales associated with cotton production (which uses just 2.3% of the world's arable land), and intensive water use, with 73% of global production currently dependent on irrigation.

While many smallholder cotton farmers are driven into debt by the cost of pesticides and fertilisers, sustainable cotton production has the potential to lift farmers out of poverty by providing a more stable income and improving working conditions.- Solidaridad Network, January 2017

We have also learned while working with farmers in different parts of the world that from the earlier trend of using just one or two 'new' biopesticides (alongwith the conventional agrochemicals) the demand for more products and even 'total' packages are increasing.

NEED FOR GLOBAL CO-OPERATION

In late 1980s we signed a joint R & D project on Biopesticide Development between GSFC – R & D Centre – India and Ecogen – USA. A handsome grant was provided to both our teams under PACT (Program for Accelerated Commercial Technology) under ICICI and India and US Govt. This was perhaps first such collaborative project of its kind and a great learning experience for both the teams.Ecogen was established as one of the first generation dedicated Biopesticides companies in USA and while leading the R & D programs at GSFC, I had put Biopesticides as one of the most important areas.

The joint project greatly realized the potential, limitations and challenges for two very different agricultural scenes. But it also helped appreciating the path ahead and the approach required for commercialization. It also becomes clear that the challenges for Biopesticide industries are global and converting the new knowledge to practical farm practices benefit a great deal with mutual cooperation. Though the initial focus was for Bt (Bacillus Thurigiensis) in the joint program, as per the need of time several Biofungicides and Bionematicides were added.

The emerging opportunities and challenges require close cooperation and knowledge transfer between various groups of scientists and industries in this sector. International trade is bound to be a crucial component of food security.

GLOBAL BIOPESTICIDE INDUSTRIES

From a humble beginning, Biopesticide Industries are emerging globally as one of the most important segment for sustainable farming. There are clear trends of appreciation of their importance by large chemical pesticide industries world wide. FCI (Farm Chemicals International) acknowledging their role published a special feature 'All Eyes on Bio Pesticides!' (FCI, August 2012). The overall scenario can be summarized as follows:

- Biopesticide industries are rising at a very high rate, though its size and contribution on a global scale is much smaller than chemical pesticides.
- The ability to provide alternative modes of action to traditional products which makes them a critical component in most IPM programs;
- Registration in less time than conventional chemical products because biopesticides exhibit minimal environment and humans;
- The ability to extend the life of conventional chemicals by providing resistance management benefit programs;

- Exemption from tolerances, such as reduced pre-harvest restrictions and application in environment which permits biopesticides that have no Maximum Residue Levels (MRLs) to be used on crops in urban settings;
- A high degree of worker safety and the shortest recently intervals allowed by law;
- Value-added benefits, such as improved plant health, yields and quality and an increase in beneficial and organic cultivation programs.
- The global market for biopesticide was $ 1.6 Billion in 2009 and rose to $ 3.3 Billion in 2014 i.e. 15.6% growth rate as per the study by BCC Research of USA. That growth rate was much higher than the expected rate of 3% for the synthetic pesticide industry. However, the volume of chemical pesticide industry is several times larger than biopesticide industry.
- The second important feature of biopesticide industry is that mostly it is in the SME segment. This is a good news for a balanced economic growth for the world. It also emphasizes tremendous challenge for R&D and new development as also organizing the industry for registration and global marketing network. One likely trend will be international cooperation and networking for creating win-win situation.

The estimated growth rate as mentioned earlier can significantly change in the coming years when we look at the major sectors in the future :

- Organic farming
- Rain-fed Agriculture

Biopesticides which are certified by international certifying agency like Ecocert and others, can have a substantial contribution in these two areas in the time to come.

Sustainable Agriculture & Biopesticides

The increasing emphasis on Sustainable Agriculture is the best thing to happen to biopesticides:

- Big food companies and the consumer awareness and preferences are the two factors which will increasingly pull the biopesticide growth.
- Over the past few years, the range of biopesticide products have been going up. Companies with just 2 or 3 products earlier have introduced more and more new products and formulations.
- Global organic food is US$ 40 billion now and growing at 10% . Here too biopesticides can offer US$ 2 billion market (even if 5% use) by 2015.

- Biopesticide companies will have to be more responsible in making claims.
- The bottom line is, the products have to perform and demonstrate effectiveness in the local conditions
- Standardization and quality control is the key growth driver for biopesticide industries.
- Bio pesticide will play a key role in Eco Agriculture Revolution!

REFERENCES

1. Mehta M.H. 2011. Presidential Address – Special Forum, Global Bio Pesticide Industries, ICOB-6, Chiang Mai, Thailand.
2. Frabotta David 2009. Farm Chemicals International.
3. K. Chandra and S. Greep 2005. Bio Control Agents & Bio Pesticides - RC of Bangalore Publications.
4. Pesticide Rich Food, Down To Earth, February 2011.
5. S. Jayraj 2002. The Hindu Survey of Indian Agriculture.
6. Mehta M.H. 2008. Key Note Address, First International Agri Business Conference, Mauritius, 25th September 2008.
7. Chemical Pesticide Ban, 2017. Times of India, 15th January 2017.
8. Ahlawat RPS, Arjun Mehta and M.H. Mehta 2009. Theomatic Area Speech – Linking the Stake holders - Key issues and experience in developing and commercializing Bio Pesticides – 5th International Conference on Bio Pesticide, New Delhi.
9. Ecogen – GSFC Joint Report – Bio Pesticides PACT Project 1992.
10. Frebotto David – Editor – "All Eyes on Bio Pesticides", FCI, August 2012.

"There is no waste material; only the right thing in the wrong place".

8

Bio Composting – Theory and Practice – Key to Sustainable Farming

Composting is organic matter that has been decomposed and recycled as fertilizer and soil conditioner. It is a natural process of decomposition and recycling of organic material into humus rich soil amendment.

At the simplest level, the process of composting requires making a heap of organic material sprinkled with water which break down to humus after a period of weeks or months. However, modern bio composting is a multi-step, closely monitored process with measured inputs of water, air and carbon and nitrogen rich (in proper C:N ratio) materials.

The term composting is used and understand world wide with different meanings. (I remember a funny incidence in China when during my talk I mentioned that the process of bio composting does not appear to be understood by more than 80% people. Objecting to this, the President of the academy got up and said "80% is a wrong figure! More than 95% people do not understand it !!"). For many people, composting is used to refer to several types of biological process often calling it "anaerobic digestion". The microbes used and the process involved are quite different between composting and anaerobic digestion.

INGREDIENTS FOR BIO COMPOSTING

Bio Composting organisms require four important ingredient to function well :

i) **Carbon:** for energy : the microbial oxidation of carbon produce the heat.

ii) **Nitrogen:** to grow and reproduce more organisms to oxidize carbon.

iii) **Oxygen:** For oxidizing the carbon, the decomposition process.

iv) **Water:** to maintain activity without causing anaerobic conditions.

Bio Composting Process

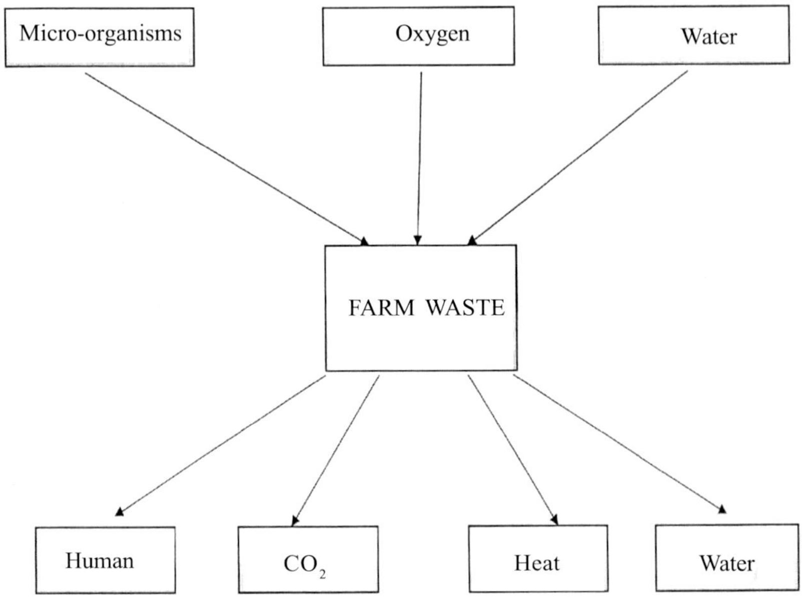

Micro-organisms
Oxygen
Water

FARM WASTE

Human
CO_2
Heat
Water

Fig. 1

The most bio composting is at Carbon : Nitrogen (C/N) ratio of 10:1 to 20:1. Rapid bio composting is favored by having C/N ratio of less than 30. Further, above C/N of 30, the subtract is nitrogen starved, and below 15, it is likely to outgas a portion of nitrogen as ammonia. In Table - 1 The optimum values of various parameters for composting are listed.

Table 1: Optimum Parameters

Parameter	Optimum Value For Composting
C:N ratio of feed	25 to 35
Particle size	10 mm for agitated systems and forced aeration, 50 mm for long heaps and natural aeration.
Moisture content	50 to 60 percent (higher values possible when using bulking agents).
Air flow	0.6 to 1.8m^3 air/day/kg volatile solids during thermophilic stage, or maintain oxygen level at 10 to 18%
Temperature	55 to 60^{0}C held for 3 days
Agitation	No agitation to periodic turning in simple systems and short bursts of vigorous agitation in mechanized systems.
pH control	Normally not necessary
Heap size	Any length, 1.5 m high and 2.5 m wide for heaps using natural aeration. With forced aeration, heap size depends on need to avoid overheating.
Activators	Use of efficient cellulolytic fungi and biofertilizers.

Ref : Tandon HPS

Too much air or water or carbon slows down the process of bio composting though during the initial process, as the pile gets hotter (upto 50 – 70%), more air and water are needed to maintain temperature till the materials are broken down.

Sequence of Decomposting

The bio composting process is dependent on micro-organisms to break down organic matter into compost. The main type of microorganisms found in active biocomposts are :

Bacteria- The most numerous of all the microorganisms found in compost. Depending on the phase of composting, mesophilic or thermophilic bacteria may predominate.

- Actinobacteria: Necessary for breaking down paper products such as newspaper, bark, etc.
- Fungi: Molds and yeast help break down materials that bacteria cannot, especially lignin in woody material.
- Protozoa: Help consume bacteria, fungi and micro organic particulates.
- Rotifers- Rotifers help control populations of bacteria and small protozoans.

Table 2:

Sequence of Decomposting
• Psychrophiles – at 15°C predominating heating process
• Mesophiles – at 15-45°C take over as psychrophiles die off
• Thermophiles – at 40-70°C consuming many bacterial

As shown in Table - 2, the three phases of bio composting are given below :

a) An initial, mesophilic phase, in which the decomposition is carried out under moderate temperatures by mesophilic microorganisms.

b) As the temperature rises, a second, thermophilic phase starts, in which the decomposition is carried out by various thermophilic bacteria under high temperatures.

c) As the supply of high-energy compounds dwindles, the temperature starts to decrease, and the mesophiles once again predominate in the maturation phase.

Benefits of Bio Compost

The use of bio compost – either alone or in combination with inorganic fertilizers is a major input for sustainable crop production. The benefits are :

- As the most important component of nutrient management.
- Mature bio compost quickly comes in equilibrium with the soil.
- Improves soil tilt and structure
- Increases water holding capacity of the soil.
- Improves aeration.
- Provides humus or organic matter, hormones and plant enzymes.
- Kills pathogenic organisms and weeds.
- Enriches organic carbon.
- Acts as buffer to changes in soil pH.
- Improved supply of micronutrients.
- It gives big savings and increases self-reliance of farmers.

Decline in soil organic carbon is a major problem in reduced productivity and bio composts play a major role in restoring soil organic matter.

Composting Technology

Apart from traditional compost pile, various approaches have been developed to handle different composting processes, locations and applications. For example:

- At the household level : Composting toilet, container composting, vermi composting.
- At the industrial level : Aerated static pile composting, -windrow composting etc.

Vermi compost uses various species of norms like earthworms, red migglers, white worms etc. to creat a heterogeneous mixture of decomposing vegetable and food wastes. The end product of break down of organic matter by species of earthworm. In India, many NGOs have established vermi compost units using a earth worm variety 'Eisenia Foetida'. There are opposite views by people like Subhash Palekar (in the book - "The Philosophy of Spiritual Farming") against this with the arguments that they have short life, limited effect and harmful accumulation of heavy metals! Several practical limitations are also reported by organizations like SEWA (Self Employer Women Association – Ahmedabad) in operation of vermi compost units.

Other methods like Bokashi in Japan (Bokashi means 'shading off' or 'Gradation') is based on practice of covering food waste with rich local soil that contain microorganism that would ferment the waste.

A composting toilet collects straw and other carbon rich materials, where pathogens are destroyed and the excreta is processed in situ and a second step composting is done.

Slow Composting to Rapid Bio Composting

A lack of healthy micro organisms community (alongwith factors like lack of oxygen, nutrients and water) is the main reason why composting processes are slow.

In a typical traditional method, composting is a slow process, requiring 3 – 4 months before the wastes are totally decomposed. The rapid method of Bio Compost or Bio Enriched method employs both fungus activator and nitrogen fixing and P – solubilizing micro organisms. Not only the time is shortened to 2 – 3 weeks but a bio compost of much better quality and value is made.

Table 3:

Consortia:

Consortia Culture:

Azotobacter, *Azospirillium*, *Torulaspora globosa*, *Bacillus coagulans*, *Celullomonas* sps., *Pleurotus* Sps., *etc.* Functions :

- Decomposition
- N-fixation
- P-solubilisation
- Trichoderma sps - Anti-fungal
- Plant Hormones
- Plant Growth Promoters
- Amino acids

In Table – 3, a list of typical 'Consortia Culture' is shown. The combination of microbes substantially helps in improving the performance of first stage of speeding up the decomposition and in the next phase enrichment of the pathogen free biomass

The media development is the key for consortia preparation and once successful, such formulations can be very effective. The next step is of scaling up. Photographs below show the activities involved.

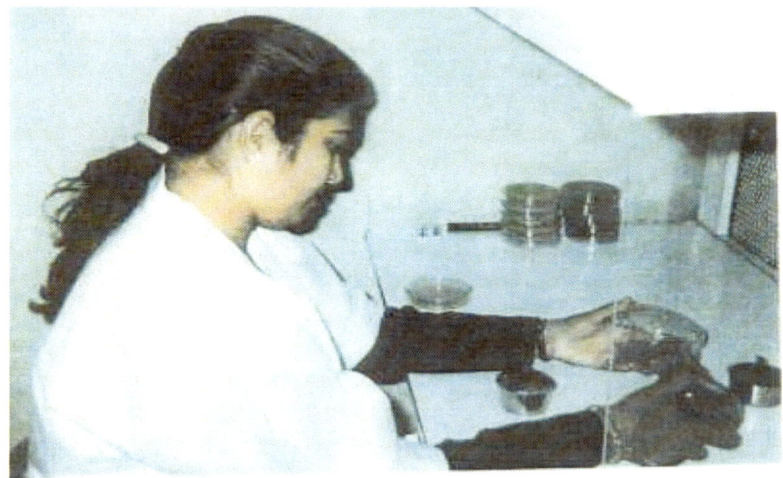

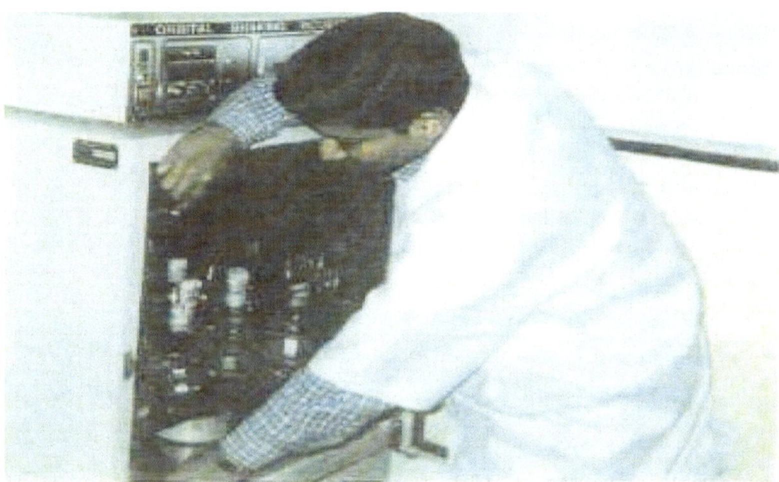

Photo – 1
Various stages of WL-D and WL-E preparation at
Modern Microbiology & Biotechnology Laboratory

Photo – 2
Latest Fermentation Technology at Gujarat Life Sciences
Scaling up from Mother Stock to Large-Scale Fermentation for Quality Production of Microbial Culture

Photo – 3
WL-D Decomposting Culture
Packed, Sealed and Ready for Despatch

Organic Manure and their compositions

Materials such as farm yard manure (FYM), animal wastes, oil cakes, rural and urban solid wastes, green manures etc are useful and time tested for improving fertility and productivity of soils. In Table – 4. The nutrient contents of such raw organic waste materials are listed.

Table 4: Nutrient Contets

Materials	Content		
	Nitrogen (N)	Phosphate (P_2O_5)	Potash (K_2O)
Cattle dung and urine mixed	0.60	0.15	0.45
Horse dung	0.70	0.25	0.55
Poultry manure (fresh)	1.0 – 1.8	1.4 – 1.8	0.8 – 0.9
Sewage sludge, dry	2.0 – 2.5	1.0 – 1.2	0.4 – 0.5
Sewage sludge, activated dry	5.0 – 6.5	3.0 – 3.5	0.5 – 0.7
MANURES, COMPOST ETC.			
Rural compost (dry)	0.5 – 1.0	0.4 – 0.8	0.8 – 1.2
Urban compost(dry)	1.0 – 1.2	0.9 – 3.0	1.0 – 2.0
Farmyard manure (dry)	0.5 – 1.5	0.4 – 0.8	0.5 – 1.9
Filter press cake	1.0 – 1.5	14.0 – 0.5	2.0 – 7.0
PLANT RESIDUES			
Rice hulls	3.0 – 0.5	0.2 – 0.5	0.3 – 0.5
Groundnut husks	1.6 – 1.8	0.3 – 0.5	1.0 – 1.7
STRAW AND STALKS			
Banana dry	0.61	0.12	1.00
Cotton	0.44	0.10	0.66
Sugarcane Trash	0.35	0.10	0.60
GREEN MANURE, FRESH			
Cowpea	0.70	0.10 – 0.20	0.60
Guar	0.34		

Ref. : Fertilizer Statistics 2003 – 04 and Handbook on Fertilizer Usages, 1994, Fertilizer Association of India, New Delhi.

The major handicap of all bulk manures are their low nutrient content and large volume. Though all farmers know the value of good composts, often the quality of composts are not proper. Strains of microorganisms which can hasten the process of composting can considerably help overcome these practices.

Typical Examples : Two Step Actions – Wonderlife – 1 and Wonderlife 2 – 3

The two step Bio Compost making process involves the consortia of decomposing micro organisms (Wonderlife 1) followed by another (Wonderlife 2 – 3) which are enriching bio fertilizer mixture. The list is given in Table – 5.

Table 5:
Consortia of Micro Organisms
(Wonderlife – 1 and Wonderlife – 2,3)
(a) WONDERLIFE – 1 (For Decomposition)

Sr. no.	Name of microorganism	Function
1.	*Azotobacter chrococcum*	Helps in nitrogen fixation
2.	*Azospirillum brasilense*	Helps in nitrogen fixation
3.	*Bacillus coagulans*	Phosphate solubilizer
4.	*Torulaspora globosa*	As a biocontrol agent
5.	*Cellulomonas cartae*	Degradation of complex carbohydrates & proteins
6.	*Pluerotus ostreatus*	Degradation of polyphenols, aminophenols, aromatic & aliphatic amines.

(b) WONDERLIFE 2, 3 (For Enrichment)

Sr. no.	Name of microorganism	Function
1.	*Azotobacter chrococcum*	Helps in nitrogen fixation
2.	*Azospirillum brasilense*	Helps in nitrogen fixation
3.	*Bacillus coagulans*	Phosphate solubilizer
4.	*Torulaspora globosa*	As a biocontrol agent
5.	*Trichoderma viridae*	Produces chitinase which helps in chitin degradation.

Steps 1 and 2 are given in Table 6

Table 6:

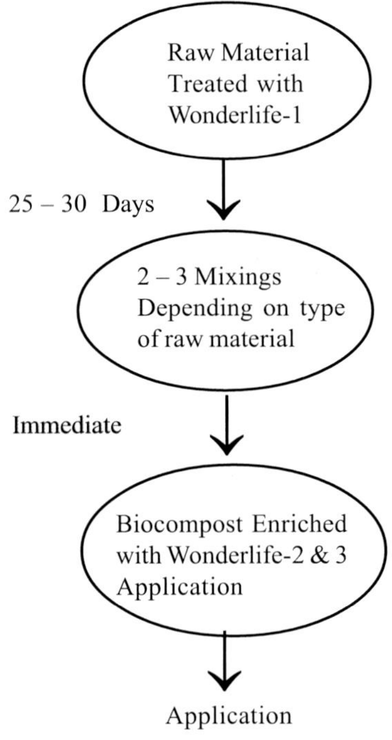

Wonderlife: 1-2-3

- Library of microbes and custom made formulations
- Any organic waste & industrial waste, separately and/or in combination
- Systematic decomposition, Enrichment & Enhancement
- Quick & Complete composting
- Excellent agricultural input
- "Better than all composts"
- Enrichment to all composts

A) Sugar Industry Waste to Bio Compost

Press mud from Sugar factories is considered to be a 'Waste Material' with great potential value. For sugar mills this waste material is a bad smelling headache but a potential source of rich bio compost. The sequence of application of consortia for a large Sugar factory in North India are shown in the enclosed photographs. Starting from day 1 of waste (press mud) dumping, formation of heaps or windrows to spraying of Wonderlife – 1 and later are explained in photographs show the sequence.

A : Farm waste + Press Mud Treatment – South Africa

Waste Dumping

Heaps or Windrow formation

B : Press Mud Treatment – Sugar Mill – North India

Manual Spraying

DAY – 10-15

DAY – 10-15

Rich Microbial Growth indicating completion of composting cycle

Manual Mixing

Machine Mixing

DAY – 25

Screening of Final Compost before packaging

Bagging or Direct – As per Farmers' Choice

Despatch of Final Product

From Waste to Best!!

Good Quality Bio Compost for market

The process of decomposing as measured is indicated in the Photographs above.

The decomposing sequence, measured with temperature changes is shown in photographs below :

In-situ Temperature measurement

Table 7: Analysis of untreated and treated Press Mud Biocompost.

Tests	Untreated	Wonderlife:1-2-3 Treated
pH	8.6	6.8
Nitrogen	1.02%	2.58%
Phosphorous	2.00%	4.70%
Potassium as K	0.43%	1.48%
Sulphur as S	0.10%	0.84%
Moisture	>55%	45-50%
Trace Micronutrients	Average	High to Very High
Smell	Pungent	'Ammonical'
Microbial Counts	1.6×10^{3}	$> 1.0 \times 10^{9}$

In Table – 7, the analysis of the enriched Bio Compost is given which shows considerable improvement in quality and enrichment of the compost. Similar on going large scale applications in other Sugar Mills in Gujarat, Tamil Nadu, Mauritius and South Africa clearly show :

a) Converting waste to high value biocomposts.

b) Cost benefit for the Sugar factories.

c) Great demand for this product.

d) Sustainable farming.

A typical performance data at Navsari Agricultural University farms on different crops are given in Table 8.

Table 8: Field Evaluation Data of Wonderlife:1-2-3 Enriched Biocompost in Agricultural Crops

Crop	Yield Increase%	Savings in Agri-Inputs %	Increased Effective Profitability %
Tomato	18	30	20
Potato	20	35	22
Egg Plant	20	40	25
Chilli	15	30	20
Banana*	25	35	25
Sugarcane*	30	35	28
Pigeon Pea	25	25	20
Onion	20	30	25
Maize	30	35	26

Notes:

1. All trials with 20% reduced chemical fertilizers of the recommended dosage
2. Single application of Wonderlife:1-2-3 Enriched Biocompost as basal dose in furrow placement; except where (*) indicated
3. Effective profitability calculated on the basis of reduction of chemical inputs including fertilizers, pesticides and fungicides
4. % figures rounded up for simplification

B) Animal Wastes from Cattle Camp

Area of Kutch – North Gujarat is known for charitable trusts running small and big camps – known as Panjrapol – to take care of old and sick cows. One such typical unit has been housing nearly 300 cows has been running on charity. A small amount of their income was through selling of untreated cow dung at very low prices in an unorganized manner. The quality of cow dung was poor and often the complaints about unhygienic dung with harmful pathogens were there. At this point, the technology of applying consortia products were demonstrated. The analytical report as shown in Table 9 made the waste pathogen free and enriched with useful microorganisms.

Table 9: Suthri Panjrapole
(Cattle Camp) – Kutch (Gujarat)
Analytical Report
GLS WL:1-2-3 Enriched

Sr. No.	Microorganism	Treated Cfu/gm	Recommended/ Allowable Cfu/gm
1.	Nitrogen Fixing Bacteria	1.0×10^9	1.0×10^6
2.	Phosphate Solubilising Microbes	1.0×10^9	1.0×10^6
3.	Anaerobic Bacteria	1.9×10^3	1.0×10^5
4.	Aerobic Bacteria	1.2×10^{10}	1.0×10^5
5.	Cellulose Decomposing Bacteria	1.3×10^2	1.0×10^3
6.	Pectin Decomposing Bacteria	Not Detected	1.0×10^3
7.	Pathogenic Bacteria	Not Detected	1.0×10^2
8.	Pathogenic Fungi	Not Detected	1.0×10^2
9.	Beneficial Fungi (Spores incl.)	1.2×10^9	1.0×10^2
10.	Nematodes	Not Detected	1.0×10^2
11.	Estimated Nitrogen	2.30 %	>2.00%
12.	Estimated phosphate	2.85%	>2.50%

Observations

- pH- 6.5 to 7.0
- High percentage of N and P
- High microbial count
- Distinct smell indicating complete composting process

Adopting such a treatment quickly and treating all the animal waste started generating handsome profits for this highly useful product and practically made the cattle camp self sufficient. The organizers would often explain to others that "normally cows give two products namely milk and dung. Our cows give only dung, no milk - and still we make profits!". Similar opportunities are for small and big cities.

C) Deoiled Cakes

Oil cakes of castor, neem, karanj, palm oil etc. are rich source of natural fertilizers, Fig. 2 and 3.

Fig. 2: Successful Application of WL-1 & WL – 2 & 3 in various Agro Industrial Wastes

Table 10: Enrichment by Consortia (Wonderlife 1-2-3)

	Phosphate Enrichment	Nitrogen Enhancement	Micronutrients &Trace elements increase by WL 1-2-3
	%	%	%
De-oiled Cakes	1.45 to 4.22	0.50 to 1.85	45%
Municipal Waste	1.65 to 3.28	1 to 2.35	45
Green Waste	0.45 to 1.56	0.55 to 1.95	50%

The enrichment data of such different cakes as well as other wastes. Particularly the treatment of castor cake in India and Palm Oil cakes in Malaysia. (This is discussed in more details in the next chapter).

Fig. 3: Palm Oil Cake

Castor processing in India, now uses microbial consortia to treat castor cake to increase its value. Ishidu Oil Mill / Jayant Oil Mill and GLS introduced a product Pragati Plus. This involves *in-situ* treatment of Castor cake at the site itself and proved to be increasingly popular with farmers in India and Asian countries like South Korea, China etc.

BOX–1

Composting of City Wastes

Composting of city wastes is now a legal requirement for all urban local bodies in India, but Central and State governments have yet to perceive it as a social good requiring official nurturing. The paper describes India's current agricultural scenario, the proven benefits of Integrated Plant Nutrient Management (IPNM) and the use of city compost in drought-proofing crops and restoring degraded soils.

Our Planning Commission estimates a shortfall of 6 million tons a year of organic manures. City Compost can fill this need and solve both problems, of man-made barrens and organic nutrient shortages. India's 35 largest cities alone can provide 5.7 million tons a year of organic manures if their biodegradable waste is composted and returned to the soil.

Mainly because of our national need for returning nutrients to our soils, and to take advantage of our tropical climate, our Municipal Solid Waste (Management & Handling) Rules 2000 [MSW Rules] require that in all ULBs, "biodegradable wastes shall be processed by composting, vermi-composting, anaerobic digestion or any other appropriate biological processing for stabilization of wastes".

Adopting Integrated Plant Nutrient Management (IPNM) for use of valuable city compost along with synthetic fertilizers will generate enormous national savings for the prosperity of India's farmers as well as the cleaning of urban India. There is scarcely any other national programme which can bring such huge benefits to both urban and rural sectors, and address the desperate need to save India's soils and sustain their productivity.

Almitra Patel et al in 'Supreme Court Committee' – Ref. 8

BIO COMPOSTS – WASTE TO WEALTH

There are several examples of treatment of small and big sewage sludge, village level wastes and agro industrial wastes, kitchen wastes etc. In fact, standard Bio Compost preparation from village wastes is discussed later in the chapter Eco Village – Clean Village. Also the post harvest wastes which are often burnt by the farmers is another attractive raw material for Bio Compost and will be discussed in the subsequent chapter.

It has been seen in a number of situations and under different agro climatic conditions that the proper use and application of good Bio Compost becomes the main and perhaps the most important input for higher and sustainable production of different crops both for conventional as well as organic farming practice. In fact, organic farming cannot be practical without the proper management and application of Bio Composts. Scientific Bio Composting with multi microbial consortia technology should become standard practice for preparing most valuable agri input for sustainable farming.

We need to standardize scientific Bio Compost making and multiply thousands of folds. Benefits of recycling Organic Waste must reconcile the social and ethical function of abolishing hunger.

There also needs to be a level playing field for different waste-processing options. Today, Waste-To-Energy (WTE) is being aggressively promoted as an alternative to composting by the Ministry of Non-Conventional Energy Sources (MNES), despite a long string of failed plants and promises and several scams engendered by its massive subsidies. Investment in WTE for say 100 tons a day of city garbage costs upto 10 times more than a compost plant of the same input capacity. The capital cost for WTE power generated is, per Megawatt, 2.5 to 3 times more than that per MW from hydel or thermal sources. Yet the pressure, from both foreign firms and international aid agencies, for promoting WTE tech technologies that are being phased out abroad is tremendous and a serious impediment to rapid decision-making by cities in favour of the statutory and far more viable composting option.

Finally, composting needs to be seen by all decision-makers as not just one of many options for processing and disposal of city waste, but an absolute imperative for nutrient recycling and soil improvement in a largely agricultural economy.

REFERENCES

1. National Institute of Molecular Biology and Bio Fertilizer University of Philippines, Los Banos, 1985, Teodula M. Corton and Santiago R. Obien, Compost production guidelines
2. Mehta M.H. Bio Composting Theory & Practices, Key Note at International Composting Conference, Beijing, 22 – 26 Oct. 2015.
3 Subhash Palekar, The Philosophy of Spiritual Farming, Amravati, Maharashtra, July 2010.
4. Personal Discussion with Women Volunteers of SEWA, Ahmedabad, Volunteers and Farmers – Suthri Panjarapol, Kutch, Sugar Industry, South Africa and India
5. GLS – TSA – NAU (Navsari Agri University, Experimental Farm) Report, October 2007.
6. Handbook of Agriculture, ICAR, New Delhi, Rev. Edition, January 2006.
7. Transforming Rural India – Amrita Self – Reliant Village - A – Sedve – Mata Amrita Mayi Math, 2013.
8. Almitra Patel, Policies to encourage use of city compost in India; TERI – Green India, 2001.
9. Tandon H.L.S, Recycling of Wastes in Agriculture FDCO; New Delhi, 1995.

9

Crop Residue Burning Problem and Bio Composting

Large quantities of crop residues are produced and if properly used, they can provide cost effective means for supplying organic matter, plant nutrients and improved soil health. The highly lignocellulosic nature of crop residues indicate their potential role in maintaining soil organic matter and enriching the soils with N-rich microbial biomass resulting from decompositions. However, their indiscriminate handling often burning leads to decline in soil health, air pollution and overall loss. Since most crop residues find use as feed and fuel, efficient recycling of the remaining amount can have very significant and positive effect.

Table 1: Potential of farm residues and plant nutrients in them

Crop	Grain: straw ratio	Residue Production(ml)	Percent (oven dry basis)		
			N	P_2O_5	K_2O
Rice straw	1:1.5	106.01	0.58	0.23	1.66
Wheat straw	1:1.5	80.99	0.49	0.25	1.28
Sorghum	1:2.0	21.04	0.40	0.23	2.17
Pearl Millet	1:2.0	15.58	0.65	0.75	2.50
Maize	1:1.5	12.50	0.59	0.31	1.31
Pigeon pea	1:2.5	6.65	1.10	0.58	1.28
Sugarcane	1:0.2	40.92	0.35	0.04	0.50

Source: Ministry of Finance 1990-91 and FAO Field document.

The data regarding five major crops are shown. Even after use of some of them as animal feed they offer a great opportunity of using them back in to the soils which are hungry for organic carbon and other matters.

PRACTICE OF CROP RESIDUE BURNING

The problem of collection and transport of crop residue and the short time between harvesting of one crop and making the farm free for the second crop has led to the harmful practices of crop residue burning. If instead of burning it,

Crop Residue Burning

the residues can be incorporated into soils, a host of heterotrophic microorganisms bring about minor discharge of carbon and other elements contained in plant residues. A large increases in beneficial soil microbial populations and organic carbon can be achieved.

Between harvesting of paddy and subsequent sowing of wheat, farmers get just 15 – 20 days. To clear the paddy stubble and to ensure timely sowing of wheat, farmers burn the paddy stubble as it involves no cost. For example, every year in Punjab, about 7 to 8 million metric tones of paddy straw is burnt openly between October and November. It is estimated that burning of one tone of rice straw accounts for loss of 5.5 Kg Nitrogen, 2.3 Kg phosphorus, 25 Kg Potasium and 1.2 Kg Sulphur besides, Organic Carbon, Crop Residues burning is a potential source of Green House Gases (GHGs) such as CH, CO, N_2O, NO_x causing global warming. Besides, burning of crop residue also emits large amount of toxic substances that are known or suspected carcinogens and could lead to various air borne/lung diseases.

The National Green Tribunal has asked Delhi and four northern states namely Haryana, Punjab, Rajasthan and Uttar Pradesh to issue stringent notifications to paddy farmers to stop burning crops, Ministry of Agriculture in consultation with ICAR has issued Guidelines to prevent crop residue burning in the country.

Sarkar et al have shown that crop residue in Panjab, Haryana and other nearby states in the rice – wheat rotation system is the main reason for smog and air pollution in cities like Delhi. Smog and air pollution (including several highly toxic and carcinogenic compounds like Benzenides, methyl cynide etc.) play havoc with the quality of air and health of soil and people !

On the other hand, average 25% of N and P, 50% S and 75% K is retained in the cereal residues making them valuable nutrient sources.

Two ways of converting such residues to useful products are discussed here :

1. Using microbial consortia :

   In this method, rapid composting of crop residue is done.

2. In-situ incorporation of residue into soil :

This is done through adoption of conservation in agriculture practice to prevent loss of moisture in soil. ICAR (Indian Council of Agriculture Research) has launched a platform project on 'Conservation' Agriculture to save cost of fuel and labour and to ensure timely planting of Kharif and rabi season crops and also to help save irrigation water retaining in the straw as surface mulching. In this, popularizing farm machines like happy seeder / suredder, turboseedsers etc. is necessary.

The mesophilic lingo-cellulolytic organisms namely fungi such as *Rhizomucor pusillus*, *Aspergillus heteromorphus*, *Aspergillus terrus*, *Aspergillus flavus*, *Trichoderma viridi*, *Chetonium* spp. and bacteria such as *Bacillus polymixa* and *Pseudomonas striata* are the promising bio consortia that enhance the decomposition process and shorten the usual period of 6–8 months by farmers practice to 1-2 months (scientific method). The fungal cultures are added @500 g mycelia mat/tone of materials and bacterial culture added @ 10 liter/tone of materials dry weight basis. Some epigeic earth worms have been identified (*Eisenia fetida*, *Eudrillus eugenia* and *Perionyx excavatus*) and inoculated them on the basis of 1 Kg /100 Kg materials and this shortened the period of decomposition to about 1.5 to 2.5 months.

BOX – 1

Composting Organic Material with high lignin content

A technique for making good compost from hand plant materials (like saw dust, pine needles, dry fallen leaves, coir mix etc.) involves mixing lime in a ratio of 5 kg per 1000 kg waste material. Liming enhanced humidification process of plant residues by enhancing microbial population and activity by weakening liquin structure. It also improves the humus quality by changing the ratio of humic to fulvic acid. Instead of lime powdered phosphate rock can be used in a ratio of 20 Kg per 1000 Kg organic waste.

Enrichment :

Farm waste in poor : N P content (0.4 - 0.8) so adding P makes the compost more balanced and supplies nutrients to microorganism. Addition or N fixing and P solubilizing cultures and secondary incubation of Azotobacter, *Azospirillium lipoferum* and *Azospirillum brasilence* (N fixers) and *Bacillus megaterium* or *Pseudomonas* or (P solubiliser) can be Sprinkled when decomposed material is about 1 month old.

Intensive studies at The Science Ashram / GLS labs have been taken up for rice and wheat straw and sugar cane trash. The practical need of the farmers requires that the Bio Compost material is made ready in the shortest possible time as he can wait for more than 20 – 25 days before the next sowing. This is done through two step process as shown in Fig – 1,2 and 3.

Fig. 1: Consortia

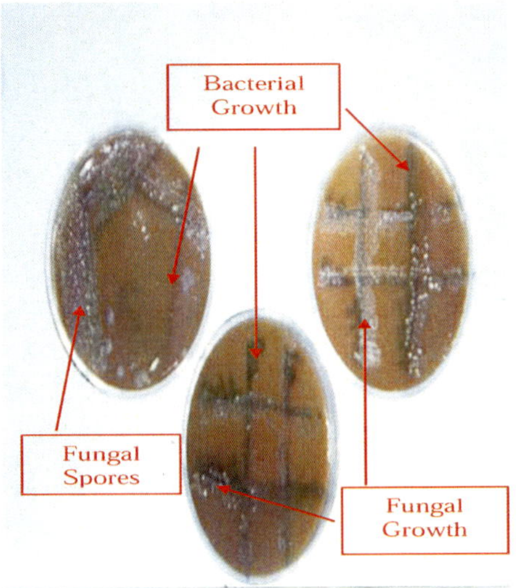

Fig. 2:

Process Sequence for application of Formulation,1 (Relife) and Formulation, 2 (Wonderlife 2 and 3).

Step 1: Application of Formulation, 1 (Relife) for Biocomposting of Agricultural Waste on the field (*in-situ*) after harvesting.

Cut down the grass/stems of the last crop after harvest.
Shredding into smaller pieces is recommended.

↓

For treating 1 acre worth of agri-waste/residues, take 5 liters Formulation, 1 (Relife)
and
Add to 100 – 150 liters of water.
(Alternatively take 5 liters and mix with about 200 Kg soil / bentonite mixture)

↓

Spray this mixture evenly on the agri-waste.

↓

Mix up the waste with the soil by turning the soil upside down using a disc harrow.

↓

Provide irrigation.

Fig. 3:

Step 2: Application of Formulation, 2 (Wonderlife 2 and 3).

(To be applied 2 or 3 weeks after the application of Formulation, 1 above)

Mix formulation, 2 (Wonderlife 2-3) about 5 liters in 20 liters of water.

↓

Spray this in 1 acre land (hand or mechanized)

Turn the soil with hand or using disc harrow / Rotavator *etc* or hand.

Cultures of selected microorganisms and preparations of certain enzymes have been found to speed up the decomposition process and there is a good care to investigate different combinations of microbial inoculants and enzyme starters for activating the decomposition of the soil – incorporated plant residues, particularly those of lignocellulosic nature.

BANANA CORP RESIDUES

Another practical problem of Banana post harvest waste disposal was also discussed often at farmers meetings in typically intensive belts. Anand – Navsari of Gujarat and around Trichy in Tamil Nadu. Both these areas are well known for high density of Banana plantation. It was told that earlier, after the harvest of Banana fruits, the stem and other materials were simply thrown away. Later, even this became difficult due to labour shortage.

While interacting with the farmers, a need to have a system to recycle the post harvest banana waste was emphasized. The major emphasize was on using the biomass (alongwith a large quantity of water in the stems) for soil enrichment in-situ. Using multi-microbial mixture on solid carrier material as explained in Fig – 2 and 3 as mixing with soil and the bio mass as shown in pictures was carried out at a banana farm between Vadodara – Anand Banana belt.

Post Harvest Residue Treatment / Conversion
Photograph of application of Formulation 1 (Relife / Wonderlife)
Experiments at Banana Farm - Chapad (Vadodara – Gujarat) in 2015

Such a method of *in-situ* treatment is easily adoptable and has the following advantages :

- Improved soil organic carbon and nutrient
- Retention of moisture
- Saving labour cost
- Saving in input cost and external plant nutrient
- Retention of micronutrient.

BOX – 2
C/N RATIO

Apart from moisture content of the compost mixture, the other important calculation is C/N ratio.

Grass dipping and green vegetables have lower C/N ratio due to higher N.

Dried leaves, wood chips etc have low N and therefore higher C/N.

The usual C/N ratio is too low, the compost becomes too hot, killing, microbes may go aeration resulting in foul smelling.

C/N decreases during composting and will become about 10:1 at the end.

It has been observed that the nitrogen in residues left on the surface was immobilized much longer than nitrogen in the buried phytomass and the decay rates increased with amount of water applied.

The crop residues have high C/N ratios which commonly range from 50 to 150. How fast the nutrients will be released depends on the activity of microbial decomposers, which is moisture and temperature dependent. In colder climates more than half of residue may remain undecomposed even after 1 year. By contrast, in hot humid climates residues decompose rapidly, making nutrient much more readily available.

WASTES FROM PINE TREES IN THE HIMALAYAS

These are ‘hard’ plant wastes and their treatment and reuse pause important challenges.

One of the major reasons for land slides and flash floods in Himalayas has been attributed to monoculture pine plantation on the hills and the waste pine needles. The monoculture and artificial forestation on the hills killed the ethnic biodiversity of the region and developed monoculture which focused on pine forests that proved very profitable. Pine needles created smooth, dry carped which absorbs nothing and turned the soils acidic. Pine wastes not only spoil the soil but also gave rise to forest fires. If such waste needles can be converted to bio compost it can considerably help the local populations. Based on the requests from a local NGO [SEED (Science for Equity, Empowerment & Development)] and after interactions with local populations, development work for preparing Bio Composts from such agro waste was taken up.

Pine Waste to Bio Compost

The bio compost prepared can be used for local farms. The fine powder from the bio composts was also used as a useful material for preparing 'Seed Balls' (i.e. covering the seeds with appropriate mixture for three plantations in hilly areas. More details of this are discussed in the later chapter on 'Bio shield'). It has been proposed to have several such small scale Bio Compost units to be operated by village level co-operatives of women. When implemented successfully it can help convert a big problem into profitable opportunity.

PALM OIL MILL WASTES IN MALAYSIA

Palm Oil contributes nearly 1/5th of worldwide vegetable oil production and Malaysia accounts for nearly 50% of total production. Palm oil production is an integrated process with several stages starting from cultivation, oil crushing and refining etc and generate large quantities of by products and waste products. If not dealt with in a scientific manner, it could lead to the deterioration in the eco system.

All wastes from Palm Oil Mill and their common uses are listed in Table - 2.

Table 2: Various palm mill wastes and their common uses

No	Type of Waste Residue	Uses
1.	Fronds, Trunks and Leaves	• Used as mulching material in the fields which helps in moisture retention. • Also used as roofing material and some are processed as furniture.
2.	Empty Fruit Bunch(EFB)	• Earlier used for generating steam for the mills and ash residues used as fertilizer. A major cause of environmental pollution so done on low key. • As raw material for products such as paneling, composites, fine chemicals, pulp and paper as well as compost and bio-fertilizer [14] • Main substrate for the cultivation of *Pleurotus ostreatus* (oyster mushroom) (Tabi et al., 2008)[15] • Most of it is just disposed off back into the fields as the above uses are not in large scale.
3.	Palm Press Fibre(PPF)	• Fuel for the mills • Used as a substrate for animal feed in addition to soymeal, fishmeal. • Used for making fibre boards. • Polymeric composites for building materials referred to as AGROLUMBER for products like wall panels, sub-floors, doors and furniture parts. (MPOB 2009)[14] • Used as potting material for ornamental plants • Currently trials are on-going utilizing fibre+ Pome for vermicompositing.
4.	Decanter cake	• Used as animal feed • Used in combination with inorganic fertilizer to improve soil quality (Haron and Mohammed, 2008)[17] • Currently on going work using dried, powdered form of DC as biofertilizer for vegetable gardening.[18]
5.	Palm Kernel Cake(PKC)	Suitable as feedstock because it has 48% carbohydrate and 19% protein (Kolade et al., 2005).[19]
6.	Shells	• Used mainly for fuel. • Converted into activated carbon for water purification purposes.
7.	POME	• Mainly used for Irrigation purposes but due to its acidic nature is quite toxic to flora and hence needs to be treated. • Carotenes are extracted from POME by pharmaceutical industries

POME being applied onto EFB using a mobile sprayer

Combining POME (Palm Oil Effluent) and EFB (Empty Fruit Bunches) and spraying with consortia of microbes as shown in Photograph can give a systematic solution for the waste conversion to highly useful Biocompost / Bio Fertilizer. In this, the windrows of EFB are arranged and applying POME at set intervals, turning and aerating to maintain oxygen level thus aerobic decomposition and prevention of methane evolution is done. Typical values of organic carbon, plant nutrients and micronutrients are listed in Table – 3 which make the finished product as very welcome soil input.

Table 3: Nutrient composition of Biocompost

Type of Compost	OC	N	P	K	Ca	Mg	Na	Zn	Cu	Fe
Bio compost	12.2	0.8	0.38	0.48	2.27	0.57	<0.01	0.0012	0.0017	1.1690

Source: (Nagavallemma et al., 2004)[13]

CONCLUSION

Proper management of crop residues is mot important for conservation of soil organic matter, recyclation of plant nutrients, provision of optimum contributions for soil biota and prevent undesirable environmental effects. At the same time, minimizing the impacts of human activities on tropospheric chemistry requires lower emissions of green house and other gases and avoiding health hazards by smoke due to phytomass burning. Appropriate management of crop residues can help achieve many such goals.

The above some examples in different conditions show how it is possible to achieve this. Many more good examples and case studies can be listed. However, what is important is to promote these national ways of dealing with straws, stalks, leaves etc. not as undesirable left overs of much more highly valued crops – as valuable resources that provide irreplaceable and productive agroecosystems and sustainable food production.

REFERENCES

1. Sarkar S et al; Current Science, June 2013.
2. Editorial, Agri Today, December, 2014.
3. Handbook of Agriculture, ICAR, New Delhi, January,2006.
4. Mehta M.H. Solution to Crop Residue Burning and Agrowaste Treatment, Invited talk Presentation at Prime Minister's Office, New Delhi, 13th January, 2016.
5. Pimantel D, et al; 1995, Environmental and Economic of soil erosion and conservation benefits, Science 267:1117-1122.
6. Vaclav Sunil, Crop Residues – Bioscience 49 (4).
7. Jean Bonhotal, Kome Composting, Cornell Waste Management Institute, 2012.
8. A. Embrandiri et al, Palm Oil Waste Utilization, Malaysian ConsultantInternational Journal of Scientific and Research Publications, Volume 3, Issue 3, March 2013.

10

Bio Input Packages – Some Case Studies

Reducing the input costs in farming, especially agrochemicals and improving the yields of farm products in environmentally sustainable way, is one of the most important challenges. World over, the understanding of alternative eco-friendly inputs and intelligent integration of such products with the farming practices are the areas which need to be emphasized, understood and demonstrated. Our experience of working with farmers in different regions has convinced us that they all are looking for alternative, eco-friendly and cost effective inputs and the only way to convince them is practical demonstrations.

Eco safe farming is a global challenge and needs combined efforts of NGOs and farmers bodies, private and public sector industries and scientists in the institutions and universities. There are many challenges and many success stories and some typical once are listed here.

1. POTATO IN BIHAR

Bihar Government planned a major program in 2011–12 for taking up large scale organic farming. A group of experts was invited to discuss and help this initiative. After a systematic planning and joint participation of the state Government, public and private sectors a package of bio inputs was prepared. The results with respect to productivity increase were extremely encouraging as reproduced from a case study paper in Box – 1.

BOX - 1
ORGANIC SUCCESS STORY

(As published in Organic Farming Newsletter 8 (2) June 2012)

A case of Organic Potato in Bihar : World record production using Eco-friendly Agri Bi Inputs

Pushpa B. Nair, Kamlesh Gokani, Rajesh Umatt, Arjun Mehta, Rajesh Mohan, Abhishek Kumar, Sudama Mahto, Devnarayan Mahto.

Gujarat Life Sciences (P) Ltd., Vadodara-390016

Email : info@glsbiotech.com

A.R. Enterprise, Patna, Bihar :

Email : arenterprise_patna@rediffmail.com

Ecocert, Patna, Bihar

District Agriculture Officer, Nalanda

District Horticulture Officer, Nalanda

Organic Initiative of Bihar

Bihar government planned a major program in 2011-12 for taking up large scale organic farming. The chief minister of Bihar declared a major initiative by organizing International Conference on "Organic Bihar" in June 2011. Several national and international experts and practicing farmers were invited. In this invited key note address by Dr. M.H. Mehta, former Vice Chancellor of GAU (Gujarat Agricultural University) and now Chairman of "The Science Ashram/ Gujarat Life Sciences" presented a model showing how packages of Ecofriendly Agri Bio Inputs can substantially reduce the agri-input cost and at the same time improve farm productivity. Such a model of Agri Bio Inputs CERTIFIED BY France based company; ECOCERT was thought to be the most appropriate for most of the places but particularly for resource poor and backward areas.

Bihar-Nalanda District

Many parts of Bihar are relatively poor but have rich and diversified agricultural base. Again more than 70% of labor force and poor are in rural areas. It was appropriately thought that a right Ecofriendly or organic model can convert this into an opportunity.

Table 1: Eco organic package for Potato

Sr.n.	Inputs name	Per hectare/Kg/lit
1.	Cow dung	1500kg
2.	Vermicompost	100kg
3.	Poultry manure	100kg
4.	Wonderlife-G (Granules)	10kg
5.	GLS ENRICH (soil conditioner)	20kg
6.	Vamstar (root promoter)	6.25kg
7.	Tricholife (seed treatment and disease management)	2.5kg
8.	Superlife (liquid biofertilizer)	2.5 lit
9.	GLS Hot Favourate + GLS Amino mix (liquid growth promoter)	2.5 lit each
10.	GLS Kelp Extract (Growth Enzyme)	2.5 lit
11.	Bioshot (fungicide)	2.5kg
12.	Neem A life (Botanical biopesticide)	2.5 lit
13.	Beaulife (powder biopesticide)	2.5 kg

(The farmer also had used 12kg/hectare of NPK mixture but other than that no chemical fertilizer or pesticides were used.)

BOX–2

Few selected progressive farmers of village Darveshpura planted potato crop during winter of 2011–12 and based on above package reported his inputs usage as per Table–1. Although the package used by the farmer was not 100% organic meeting certification requirement, but was very close to organic as only a very small quantity of NPK fertilizer was used.

Results and Discussions

In the village Darveshpura of Nalanda District, Bihar nearly 65 hac of land was taken up for organic cultivation of potato with the above mentioned Package (Table 1). A small but progressive farmer Mr. NitishKumar in this village was taken as typical case. He had been growing potato (and other crops) in his farm since last many years and had earlier reported a highest yield of potato of nearly 30tones/ha using chemical inputs.

With his hard work, dedication and observational precision farming with above organic package, Nitish Kumar has proved that with help of organic input package, a substantial increase in production is possible. To witness the actual yield the harvesting of potato, was undertaken under the supervision of agriculture scientist and District Agriculture and horticultural, officers and a production of 72.9 tones /ha. was reported. (Hindustan Times March 15, 2012).

The entire program was taken with the support of the State Government under 'Organic Bihar'.

1. BANANA

The Central Gujarat region has done remarkably well in Banana plantation. However, Kikani and others have proposed that as against the present average production of 60 tons/ha the potential can be as high as 120 tons/ha and several measures have been suggested based on their field experiences. These include planting tissue culture varieties like Grand nine, William, Robasta etc. minimum distance of 5 ft x 5 ft, applying Bio Fungicide (30 gms in 100 ltrs) planting in June / July and one week after the plantation applying the range of Bio inputs as given in Table - 2.

Table 2 : Application of Bio Inputs and Schedule for Banana Plantation

(A) Before Plantation :

About 30 tons/ha of good Bio Compost

10 kg/ha Trichoderma

5 kg/ha Paccilomisis (Bio Nematicide)

(B) After Planting :

(i) First week :

Bio Input	Liter/ha in 100 liter water
Azotobacter	4
Phospho bacteria	2
Potash bacteria	4
K=Humate	2
See Weed extract	4

(ii) Second and Ninth month after Planting :

Nutrients	Kg/ha
NPK (20:20:20)	140
Castor Cake	280
Amino + Humic ACID	4

(iii) Third and Seventh month after Planting:

In 200 liter water	Liter/ha
Azotobacter	8
Phospho bacteria	4
Potash bacteria	8
K=Humate	4
See Weed extract	8

Recommendation like above increasingly being adopted by some farmers, though the timings and application doses vary and often not very systematically done.

2. SPICES

A dedicated research and field application program taken up at the Spices Research Centre at Jagudan (North Gujarat) S.D. Agri University. The main reasons for switching over to bio inputs was the increasing demand by the farmers and consumers for Organic Spices. As reported by Patel et al, minimizing or totally eliminating chemical inputs and providing ecofriendly inputs was taken up in their demonstration farms as well as on the farms of some progressive farmers.

Table 3: Recommendation by Bio inputs for Spices (Spices Research Center – Jagudan, Gujarat)

S.N.	Spice Crop	Fertilizer (Kg/H)		Bio Inputs Recommended
		N	P	
1	Cumin (Jira)	40	15	Seed Coating with Azotobactor and PSB
2	Ani Seeds / Fennel Seeds (Variyali)	100	60	Bio Compost, Seed Coating with Azotobactor & PSB
3	Coriander Seeds (Dhania)	20	10	Bio Compost
4	Fenu Greek (Methi)	20	40	Seed Coating + 10 tons Compost (5 Kg. Azospirilium/ha + Rhizobium+PSB30)
5	Parsley (Ajmo)	20	20	Bio Compost + DOC

It has been noticed that more and more farmers are including Bio inputs for Spice cultivation though only relatively smaller number get into certified organic products. A special agency – Gujarat Organic Products Certification Agency (GOPCA) has been created by the State Government and the Scientists of the center are working closely to encourage farmers to take up this opportunity.

Isabgol (Psyllium) is another important spice which grows in semi-desert areas of North Gujarat and Rajasthan. The small and marginal farmers of Bikaner – Rajasthan grow Psyllium under dry rain fed conditions and most have no access to chemical fertilizers and pesticides. This has been a big advantage as every year, European agents come to Bikaner area and buy this important spice for use in medicine as an ingredient and for breakfast and other meals.

Jaimini *et al.* reported upto 30% increase in the yield of Psyllium when inputs like Bio Fertilizers, Bio Composts and Bio Fungicide was used.

There was a visible improvement in the income of the farmers and the exporter also benefited considerably.

Spicy Crops are particularly at great advantage when using eco-friendly bio inputs.

3. ORGANIC PACKAGE DEVELOPMENT – A TRIANGULAR ARRANGEMENT

In 2004, a triangular agreement was signed between Navsari Agricultural University, The Science Ashram (an NGO) and Gujarat Life Sciences (a Company) to take up preparation of 'Organic Packages' of eco-friendly inputs for a variety of crops. It was felt that such a Golden Triangle (Fig : 1) will be the best way to arrive at the recommended package of inputs.

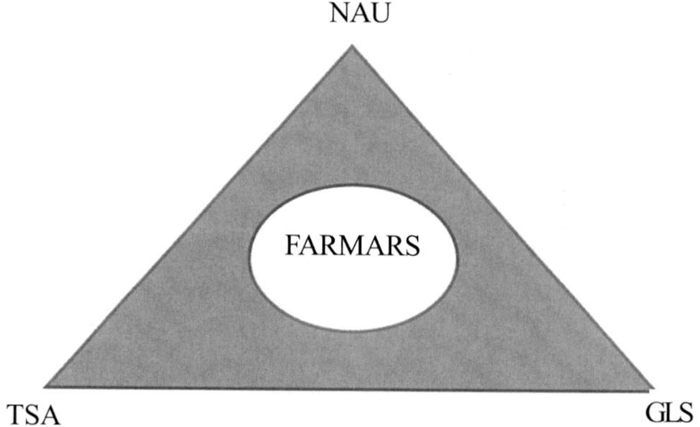

Fig. 1: The Triangular Arrangement

NAU =Navsari Agricultural University

TSA = The Science Ashram (NGO)

GLS = Gujarat Life Sciences (Company)

A broad package as shown in Table - 4 was prepared after consultation with scientists are farmers for trials on different crops.

The combination studies were carried out on Sugar Cane, Rice, Mango, Cotton and Salicornia. The average results at the end of two seasons are summarized by Ahlawat *et al.* in a joint publication as follows :

a) The average cost of inputs decreased by nearly 30%. This included the reduction in the use of DAP or Urea.

b) Average farm production increased from 18 to 30%

c) Population of natural predators like *Trichogramma*, *Chrysopoeia*, Lady bug beetle etc. increased.

Table 4: Application for experiments : NAU-GLS-TSA

Sr.	Product	Application
1.	GLS Enriched Bio-compost	• Before planting or with basal fertilizer dose
2.	Wonder life-G	• With planting / sowing. Cannot be *mixed with chemical pesticides*
3.	Super Bio Life	• Foliar spray at germination, flower setting and fruit setting stages • Spray for Lawns / Golf Courses / Floriculture
4.	Wonderlife-2,3	• For enhancement of bio-compost. Cannot be *mixed with chemicals*
5.	Tricho Life	• For fungal diseases control. Can not be cured with chemicals
6.	GLS Bioneematicide	• Nematode control
7.	Thrip –Thrash	• Thrips control
8.	Mite-No-Mite	• Mite control
9.	Mono-Shot	• Aphids, Mealy Bugs, White-flies, Sucking pest control • Also effective against Diamond Back Moth, Grubs, Borers

CONCLUSIONS

These case studies are only representative ones and in other chapters also several specific field scale results are given.

It may be emphasized that such case studies and field trial results though not very exacting in nature are very valuable. They give first hand experience and have greatly helped in building confidence among the farmers and Scientists.

REFERENCES

1. P.B. Nair et al : A case of Organic Potato in Bihar – World Record Production using Eco-friendly Agri Bio Inputs; Organic Farming News Letter – June 2012
2. Kikani K.P. : Banana Plantation – Special care – Krishi Jeevan (Gujarat), Dec. 2016
3. Patel S.M. et al: Organic Farming of Spices – Krishi Jeevan (Gujarat), March 2016
4. Jaimini S.N. and Nair PBN; Eco-friendly inputs for Organic Psyllium in Bikaner District of Rajasthan, SDAU Journal of Agronomy, Oct. 2008
5. Ahlawat R.P.S., B.N. Kalambo, AM Mehta, K. Gokani, M.H. Mehta – "Scientific Package of Practice for Organic Cultivation of various Crops'. National Conference. On Agronomy, Navsari, Nov. 2007.
6. Hindustan Times – Patna, Thurglar, March 15, 2012.

"Clean Villages lead to clean states and to clean nation. Then life transforms into a happy society".

— APJ Abdul Kalam

11

Clean Village – Eco Village and Agrowaste Utilization

Indian Government has several developmental programmes to promote agriculture and rural sector. It includes Drought Prone Area Program, Antodaya Programme, Integrated Rural Development Programme, Hill Area Development Programme, Desert Development Programme, Intensive Jawahar Rojgar Yojna and Employment Assurance Scheme, Indira AwaasYojna, Integrated Waste Land Development Programme, National Food Security Mission, National Horticulture Mission, etc. Mahatma Gandhi National Rural Employment Guarantee Scheme (MGNREGS) has the main focus to provide guaranteed employment for 100 days in a year to every rural household. Since 2014, the Swachh Bharat Abhiyan has been a major initiative by the Government of India.

THE FOOD SECURITY BILL

The Food Security Bill is here. Some of the salient features are :

Nearly 82 crore (820 m) poor people to get food grains at highly subsidized rate, nearly 20 crore poor in urban areas and 62 crore in rural areas will be eligible to get 5 kg rice, wheat and coarse grains at Rs.3, 2 and 1 respectively. Additionally for pregnant women, there will be a provision for free food for 10 months and for children below 7 years, food will be free. There is one danger point feared in this scheme that in the long run, many farmers will stop farming and go for easier way for casual labour and take food security cover in stead of farming the low production lands. But on the other hand we should look at the solution to this keeping in mind that the Food Security Bill is here and how best we can build on our strength rather than weakness. The areas of strength to look for should be :

- Agriculture and food grain will be critically most important in the overall world scenario in the coming years.
- India will have the largest population of young people, who effectively drawn to agriculture and agro industries can bring a quantum jump.
- New agri bio in puts such as biofertilizers, biopesticides, biocomposts, etc. are emerging as very effective and low cost inputs for the marginal farmers.
- In rain fed area, the contribution of livestock along with crop farming will play a very important role.
- Technologies like mKRISHI, small agro processing units such a oil mills, cottage food industries, etc. have a great potential.

EMPOWERING MARGINAL FARMERS

a) India lives in its villages. About 70% of India's population (around 833.1 Mn) live in nearly 650 thousand villages. This is the largest rural population in the world (2011 India Census). Further there are more than 550 Mn youths (below 35 years) in India today and of this number, 500 Mn youths live in rural areas. Out of the India's BPL (Below Poverty Line) population of about 300 Mn, 80% live in villages.

b) The production and productivity of small farmers need maximum emphasis for achieving food security and food availability. The rain fed agriculture is more significant for the poor countries and nearly 70% of the world staple food comes from rain fed areas. In India, almost 57% (18 Mn Ha) is rain fed contributing nearly 44% to food grain production. Globally the figures for rain fed agriculture are 93% in Subsaharan-Africa, 87% in Latin America, 67% in Near East, 65% in East Asia and 58% in South Asia.

c) Again such background and the scenario with respect to Food Security Bill, there are positive signs that we can easily increase the productivity of marginal farmers by nearly 200% through adaptation of appropriate technologies, waste management-composting, water management, agri bio inputs like biofertilizers, biopesticides and integrated farming with emphasis on livestock.

CLEANING VILLAGES PROFITABLY

We have the crisis of lowering of farm productivity. One main reason is the poor soil health and poor organic matter. SOM (Soil Organic Matter) consists of active organic matter, humus, bacteria, fungi, earthworms, nematodes, plant roots, etc. SOM and humus serve as reservoirs of living organisms which mobilizes plant nutrients. Our soils are hungry for organic carbon. Even in the green revolution areas, the low organic carbon has pulled down the productivity and fertilizer use efficiency drastically.

CONSORTIA OF MICRO-ORGANISM - PRACTICAL DEMONSTRATION

It has been possible to develop and demonstrate that a multi-microbial consortia culture consisting of various combinations of *Azotobacter*, *Azospirillum*, *Torulospora globosa*, *Bacillus coagulans*, *Celullomonas* spp, *Pleurotus* spp, *etc*. can be applied very effectively for productivity enhancement in a sustainable manner. The major functions of such a culture, when applied to agrowastes is improvement of N-fixation, P-solubilisation, plant growth enhancement and anti-fungal activity.

Fig. 1: Agrowastes to high value biocompost

In Table 1, the results of applying microbial mixtures (described in earlier chapter) clearly shows the enrichment. Further, the end products are standard and pathogen free and can become a major source of soil enrichment and plant nutrients.

Table 1: Agrowaste to Biocomposts

	Enrichment by Microbial Consortia (Wonderlife 1-2-3)		
	Phosphate Enrichment %	Nitrogen Enhancement %	Micronutrients & trace elements increase by WL 1-2-3 %
Press mud Waste	1.33 to 3.53	0.75 to 2.80	55%
FYM	0.87 to 2.50	0.87 to 2.50	40
De-oiled cakes	1.45 to 4.22	0.50 to 1.85	45%
Municipal Waste	1.65 to 3.28	1 to 2.35	45
Green Waste	0.45 to 1.56	0.55 to 1.95	50%

In another situation in Uttarakhand, the agro waste – twigs of pine trees makes an interesting case study. Already the monoculture of pine, which has commercial advantage in short term, has created a major ecological problem in Himalayan area. These wastes created a smooth dry carpet, which absorbs nothing and which at the same time, prevents the development of any undergrowth of other plants. In fact, it was proven that during the recent crisis, the ground in pine forest is bare and the soils acidic and mainly responsible for the loss of green cover and resultant problem of land slides and forest fires. Our experiments of converting such wastes into biocompost appear promising (Fig. 2) and can be taken up preferably with local women farmers of Gadhwal district in Uttarakhand

(a) Pine Agrowaste to Biocomposts

(b) Biocompost from Pine Agrowaste in Uttarkashi (Himalaya)

Fig. 2: Pine Agrowastes to Biocompost

as proposed with local NGOs, several such units can be run profitably, help improve the local farms and considerably reduce the erosion and land slide problem in the hilly areas.

AGRO WASTE USE AND INTEGRATED FARMING

Together with biocompost, a complete agri bio package of biofertilizers, biopesticides, bio growth enhancers can be offered to the marginal farmers as such a package. It is low cost, farmer-friendly and ecofriendly and a boon for sustainable agriculture. Contrary to the earlier beliefs, they have clearly shown higher productivity in many different situations such as Bihar, Rajasthan, Tamil Nadu, Kutch-Gujarat, etc.

Such bio package in a typical un-irrigated field emerged as the most cost effective and eco-friendly model. In fact, it was proven that during the crisis, the new and appropriate technologies were adopted by the small and marginal farmers quickly.

CLEAN VILLAGE – ECO VILLAGE

At the International Summit on Village Development at Amrita VishwaVidyalayam, Kollam, Kerala (25-26, Sept. 2013), there was much excitement and discussion to launch a comprehensive programme for 101 villages in different parts of the country. Naturally, eco agriculture should form a backbone of such a holistic programme, which will have other components like health & hygiene, value-added marketing, bioshields and agroforestry, education, water supply, etc.

A Clean Village – Eco Village initiative can bring about a major transformation in the rural agrarian economy. Under such a programme, agrowastes to biocomposts and complete agri-bio input can be effectively used. Such a model has been discussed at several forums, including at a major initiative taken by Amrita University.

Science and Spirituality to combine for our villages – our world
Ref. Amrita University, Publication September 2013

Similarly, a plan was discussed with the Women Volunteers of SEWA (Self Employed Women's Association) and Gujarat Vidyapith with Smt. Elaben Bhatt. A typical exercise (Box) attracted the attention of Women Entrepreneurs who have shown keen interest in taking up Eco Village program.

Demonstration under Clean Village Eco Village Program at Randheja Village, Gujarat with SEWA Women and Smt. Elaben Bhatt, Chancellor – Gujarat Vidyapith

BOX – 1

Economics of Village Bio Compost Unit (prepared for SEWA)

Waste type:

A. Cow dung/Sewage

B. Farmyard waste, Garden waste, post harvest waste and food waste etc.

C. Mixture of A and B.

Requirement:

Collection of wastes, Chopping and cutting of B type waste, Microbial consortia mixture for decomposting(Wonderlife 1) and microbial biofertilizer (Wonderlife 2,3)after decomposting of waste.

Procedure:

1. Mix 1 litre of Wonderlife 1 formulation with 100 litre of water and spray on the 1 ton of organic waste.
2. After 15 every days turn over the heap of the waste and spray the water to maintain the moisture of the waste to 30-40%
3. After 30-40 days compost will be ready.
4. Spray the 4 liter of Wonderlife 2, 3 on 1 ton of compost or put 200ml per 50kg of compost in the bag.
5. Compost can be sold in the market after enrichment with Wonderlife 2, 3 @ 250/- per bag of 50 kg.

Costing/Benefits:

1. Waste of all types free.
2. Cost of Wonderlife 1 is Rs.150 per liter.
3. Cost of 4 liter Wonderlife 2,3 is Rs.500 per liter.
4. Labour cost per ton is Rs. 900 per ton.
5. Total cost is Rs.1,550 per ton.
6. Cost of50 kg bag isRs. 16. Total cost of 1 ton (20 bags) would be Rs. 320

Selling price –Cost price =Profit per ton.

250 x 20 - 1870 = Rs. 3130 per ton.

Taken up together with integrated programs, it can bring about a major change in the lives of the people. Empowering the marginal farmers can not only solve the crisis of people living in rural areas and moving to urban areas, but also arrest the problem of food supply distribution and environmental degradation. The Clean Village – Eco Village initiative can make India a model for other countries.

CONCLUDING REMARKS

1. The pressing need is to take up sustainable programmes to empower the small and marginal farmers. This can be done through the holistic plans of agrowastes utilization alongwith low cost agri bio inputs and integrated farming.
2. Linking up Swachh Bharat – Clean Village and converting village agrowastes to quality Bio Composts can serve the dual purpose of having clean villages and improved agricultural productivity in a sustainable way.

REFERENCES

1. Sharma K.D., Current Science, June-2011
2. Mehta M.H., Clean Village – Eco Village - Agriculture Year Book 2013.
3. Clean Village program : Amrita University Publication, Kollam, Kerala, September 2013.
4. Science for Equity, Environment & Development, (SEED) Uttarkashi Report, November 2014.

"Combination of agriculture with manufacturing industries; gradual abolition of the distinction between town and country by a more equitable distribution".

— *Communist Manifesto*

12

Urban Eco Agriculture

Normally we associate our agriculture and food supply with farms and villages. The new trends of greener cities and urban and peri urban horticulture (UPH) are emerging very fast. Towns and cities in the world's developing countries are growing at great speed. Ten years ago, an estimated 40 percent of the developing world's population or 2 billion people lived in urban areas. Since then, their numbers have expanded twice as fast as total population growth, to more than 2.5 billion. By 2025, more than half the developing world's population of 3.5 billion will be urban. Again most of the world's fastest growing cities are in Asia and Africa with young population.

The concept of "Green Cities" is being discussed for environmental sustainability. However, the core principles of Greener Cities can guide urban development that ensures food security, clean environment and economic well being. Urban and peri-urban horticulture, together with eco-friendly home and public gardens and sports fields can offer a solution for a better future.

GREENER CITIES AND ECO AGRICULTURE

FAO's program for UPH is major initiative and the following advantages and plans taken from FAO's report make the reasons clear:

Urban and peri-urban horticulture (or UPH) is the cultivation of a wide range of crops - including fruit, vegetables, roots, tubers and ornamental plants - within cities and towns and in their surrounding areas. It is estimated that 130 million urban residents in Africa and 230 million in Latin America engage in agriculture, mainly horticulture, to provide food for their families or to earn income from sales.

The FAO program have demonstrated how horticulture helps empower the urban poor, and contributes to their food security and nutrition. But it can also help grow greener cities that are better able to cope with social and environmental challenges, from slum improvement and management of urban wastes to job creation and community development.

Several private initiatives of urban agriculture and horticulture are already seen around the world. The main driving force and benefits are listed here :

i) Safe, clean environment

Linking waste management to horticulture helps to keep the urban environment clean, reduce health hazards and boost production of fresh food

Pollution in rapidly expanding cities poses a serious threat to public health. Lacking adequate sewerage systems and treatment plants, many cities discharge daily huge volumes of raw human wastes and industrial effluent into the environment.

Urban and peri-urban horticulture can turn waste into a productive resource. In North America, cities routinely recycle organic waste and offer it to citizens as compost for home gardens. In Addis Ababa, a private company collects each day some 3.5 tonnes of organic waste and converts it into almost two tonnes of high-quality fertilizer. Cuba's national programme for UPH prohibits chemical fertilizer in cities and encourages instead organic composting.

UPH has other environmental benefits. It reduces the need to transport produce into cities from distant rural areas, generating fuel savings, fewer carbon dioxide emissions and less air pollution. It lowers city temperatures - in Cairo, rooftops planted with vegetables are 7° C cooler than those next door - and, when practised on greenbelts, improves landscapes and citizens' quality of life.

ii) Sustainable livelihoods

Urban and peri-urban horticulture provides livelihoods that are resilient to economic downturns and food price hikes, and contribute to cities' economic development

The International Labour Organization estimates that 180 million of the developing world's urban population are jobless and another 550 million earn just enough to survive in the informal economy. Over the next 10 years, almost 500 million people, many of them from rural areas, will enter the jobs market. Unless developing countries create more decent, productive work opportunities, the number of unemployed and working poor could reach 45 percent of their urban populations by 2020.

Urban and peri-urban horticulture offers a pathway out of poverty. It has low start-up costs, short production cycles, and high yields per unit of time and unit of land and water. Its produce has high market value. Because it is very labour intensive, horticulture creates employment for the jobless, particularly people newly arrived from rural areas.

Gardening can be profitable even on a very small scale. In Dakar, women kept 30 percent of the vegetables grown in their micro-gardens for home consumption, and sold surpluses through family kiosks, earning the equivalent of a labourer's wages. In the slum areas of Lima, women practise UPH part-time to earn extra income, and still have time for household tasks and child care.

iii) Food and nutrition security

Growing fruit and vegetables in and around cities increases the supply of fresh, nutritious produce and improves the urban poor's economic access to food

People have food security when they are able to grow enough food, or buy enough food, to meet their daily needs for an active, healthy life. In many of the 21st century's developing cities, all of those conditions of food security are threatened.

Access to nutritious food is a key dimension of food security. In Africa and Asia, urban households spend up to 50 percent of their food budgets on cheap "convenience" foods often deficient in the vitamins and minerals essential for health. One study found that vitamin A deficiency, a cause of blindness, was more severe among Dhaka slum dwellers than among even the rural poor.

Fruit and vegetables are the richest natural sources of micronutrients. But in developing countries, daily fruit and vegetable consumption is just 20-50 percent of FAO/World Health Organization (WHO) recommendations. Urban meals rich in low-cost fats and sugars are also responsible for rising levels of obesity and overweight. In India, diet-related chronic diseases, such as diabetes, are a growing health problem, and mainly in urban areas.

iv) Urban food security

China has integrated food production into urban development since the 1960s. Today, more than half of Beijing's vegetable supply comes from the city's own market gardens, and it costs less than produce trucked from more distant areas. Horticulture in and around Hanoi produces more than 150 000 tonnes of fruit and vegetables a year. In Cuba, which has promoted intensive UPH since the early 1990s, the sector accounts for 60 percent of horticultural production - and Cubans' per capita intake of fruit and vegetables exceeds the FAO/WHO recommended minimum.

v) Household food security and Home Gardens

FAO's programme for UPH also promotes home, school and community gardens, where the urban poor grow their own fruit and vegetables and earn income from the sale of surpluses. In the Plurinational State of Bolivia, FAO helped to introduce community greenhouses and micro-gardens in the municipality of El Alto, where 70 percent of residents live in poverty and 40 percent of children under five are malnourished.

School gardens are a proven means of promoting child nutrition. They familiarize children with horticulture, provide fresh fruit and vegetables for healthy school meals, help teachers develop nutrition courses and, when replicated at home, improve family nutrition as well. Over the past 10 years, FAO has provided tools, seeds and training to establish thousands of school gardens in more than 30 countries.

There are many benefits of planting your own garden at home and there is a world wide interest. Interest and spread of gardening activities and together with that the demand for eco-friendly inputs are increasing day by day. Small and big companies and nurseries supply natural and safe products. Also many home gardens produce their own Bio Composts form organic wastes form kitchen, animals and gardens. In Table 1 list of approved Bio Insecticides in USA for Organic Farming are listed.

Table 1: Insecticides approved for use in Organic Products (USA) for gardens

Active Ingredients	Origin	Pests, Insects Controller
Bacillus thuringensis (*Bt*)	A toxin produced from soil bacterium (Bt)	Caterpillar such as Cabbage, looper house worm
Diatomaceous earth	Fossilised remains of diatoms, algae with a silica based hard cell	Particularly effective in soft bodies insect also beetle, floa and ants
Insecticide Soaps	Salts of fatty acids derived from coconut and other oils	Aphids, thrips, white flies etc.
Kaolin Clay	A naturally occurring clay minerology powder	Numerous. Stops egg laying by insects.
Neem	Neem tree (*Azadi rachta* Indica)	Gypsy moths, leaf miner Whitefly, thrips, mealy bug, caterpillar etc.
Pyrethrins	*Chrysanthemum* flowers	General, beetles, caterpillar, sucking pests
Sulfur	The mineral sulfur	Spider mites, psyllias,thrips
Spinosads	Soil actinomycete (*Saccharopolyspora Spinosa*)	Numerous – Motus, caterpillar, leaf miner, potato beetles, fire ants

In view of increasing demand and interest by people engaged in home gardens such guidelines are issued by Agricultural Universities and Research Centers.

Again in Table – 2, the list of nutrient content and release notes of some organic fertilizers are given to help the enthusiastic gardeners.

Table 2: Nutrient content and release rates of some organic fertilizer

Materials	N%	P_2O_5	K_2O	Relative availability
Cattle manure	0.25	0.15	0.25	Medium
Poultry (15% water)	6	4.0	3.0	Medium – rapid
Kelp	0.9	0.5	1.0 – 4.0	Slow
Deoiled cake (Soya)	6.7	1.6	2.3	Slow
Fish meal	10	4.0	0	Slow

In another situation, during our interactions with the city gardens in Mauritius (Africa) and the city of Vadodara (India), a great interest was shown by the respective Municipal Corporations for eco-friendly gardens. Among other things, nursery mixtures using bio compost made from agro wastes alongwith bio fungicide, soil improvers and growth enhancers were prepared as the first step.

Taking soil, sand and bio compost in the ratio of 1:1:1 and mixing a small amount of Bio inputs gave remarkably improved result for nursery plants.

Table 3: Effect of Microbial Formulations and Growth of Simarouba glauca

S.No.	Treatment	Germination (%)	Seeedling HT (CM)
1.	T_1	42	9.64
2.	T_2	60	10.20
3.	T_3	78	11.66

T_1 : Sand, Soil and FYM in the ratio of 1:1:1
T_2 : T1 + Superlife
T_3 : T1 + Superlife + Bioneematicide
Ref : Nair PBN and Jaimini S.
The Science Ashram
Vadodara

For home gardens and city parks and public gardens, this trend should increase substantially. Also for the private nurseries which supply flowers and plants to city garden lovers there is interest and opportunity. Many forestry plant nurseries are maintained by Government Forest Departments in the outskirts of cities who distribute large number of plants during monsoon months and should take up such nursery mixtures.

VERTICAL AGRICULTURE

It is estimated that by 2050 nearly 80 percent of earth population will reside in urban centers and development of vertical farming will play a key role. Among the several attractive features of vertical farming being mentioned are saving in space and water requirement.

For example, conventionally one acre of lettucerequires about 1800 – 3500 m^3/ water where as in vertical farms need only 14m^3 i.e. 1/240th as calculated by

Vertical Garden

Scientists of University of Waterloo. The design also proposes use of compost from city waste and special architecture of the building and elaborate arrangement of grow lamps, dehumidifier, water pumps, growth enhancers etc. There are some apprehensions about such sophisticated systems but these new developments will also be the areas to watch for future farming.

CONCLUSIONS

- As per FAO, Urban Agriculture is practiced by 800 million people worldwide.

- It helps low income urban residents save money on food purchases.
- Urban agri carries health and environmental risk – potential use of contaminated food and water and use of pesticides and raw organic manure can leak into water resources. These issues need proper attention. Eco Agricultural inputs will therefore be of greaterrelevance and are gaining ground.
- The actual successful demonstrations of use and application of agri bio inputs in different parts of the world suggest an urgent need to take up eco agriculture practices in urban areas in a big way.

REFERENCES

1. Alfes Lloyd, Tree Huggers June 8, 2012
2. FAO Report – Growing Greener Cities – Urban and Peri-Urban Horticulture
3. Krista Jacobson, Organic Gardening, Kentuky Master Gardener Manual, 2012
4. Eco Solutions – Brochures, New Age Co. & Gujarat Life Sciences, Vadodara, 2011
5. Dickson Despomier, The Vertical Farming - Amazon.com # 119884

13

Eco-Friendly Golf Courses and Cricket Fields

Since many decades environmentalists have warned of perils of chemicalised sports grounds, golf courses and cricket fields. They warn that our health is inextricably linked to the widespread use of toxic chemicals and also complain about the non-availability of safe alternatives.

Eco-sustainable sports fields must use the methods :

- That do not pollute
- That are environment-friendly and effective
- That economize on the use of water and energy

Organizations like ECOCERT have prepared guidelines for Eco-sustainable fields including guidelines to take care of soils and turf, use of water including recyclation and use of only eco-friendly and certified compost and organic fertilizers.

Typical Golf Courses are heavy users of pesticides and water. UNESCO World Water Development report found that an 18 hole golf course can use as much as 2.3 million liters of water every day. With the increasing awareness and concern about health and environment, the demand for eco-friendly sports fields are on the rise.

Phil Weaver – Chairman of the Professional Golfer Association (PGA) has stated that "we are keen to have a more environmental approach to the golf industry. I am not sure organic is totally achievable but it is a great aim". A numerous environmental certification system have been set up for golf courses that lead the way in terms of chemical reduction there are many limitations and difficulties. Some courses are making great sides, with recycling systems, minimum or no chemical use and alternative bio inputs. In Box – 1 & 2 such examples are given which are typical.

BOX – 1
ORGANIC GOLFS

Typical golf courses are not only heavy users of pesticides, they are also held up as the standard for suburban lawns. The goal of the Organic Golf project is to prove that golf courses can be maintained organically, and thereby demonstrate that all turf can be maintained without chemical pesticides.

A *Newsday* article from 2003 reported on an agreement between a golf course developer and environmental groups including the Neighborhood Network:

Building Plan Out of Rough - Golf deal struck at Friar's Head (*Newsday*, July 1, 2003)

"After more than three years of fighting - a battle that kept one of Long Island's most impressive golf courses from opening - developers and environmentalists reached a truce over Traditional Links at Friar's Head.

...

the two sides agreed on things such as limiting the use of pesticides, using organic fertilizers to build up the soil and keeping houses out of the heart of the unspoiled woods."

This article and the editorial "Golfers, Environmentalists Find Course Up to Par" (*Newsday*, July 7, 2003) highlight the Neighborhood Network's efforts in partnership with the Long Island Pine Barrens Society and the North Fork Environmental Council to work with the developers of the new Traditional Links Golf Course to reach a compromise which resulted in the construction of a golf course with a turf management plan which emphasizes the use of organic methods and materials in favor of chemical pesticides and synthetic fertilizers. The editorial board specifically recognized the contributions of the Neighborhood Network, North Fork Environmental Council, and the Long Island Pine Barrens Society, stating that "this outcome at least shows that, if environmentalists and business folks talk, they can reach real solutions."

The Neighborhood Network has been contacted by activist working in other states, from Massachusetts to Hawaii, for information about the Organic Golf project, and ways to eliminate pesticides from golf courses. We are planning to respond to the many inquiries received from off Long Island by developing a national clearinghouse for information related to organic golf.

Ref. : Newsday, July 2003

BOX – 2

BRITAIN'S FIRST ORGANIC GOLF COURSE

With their diamond-patterned jumpers, neatly pressed slacks and expensive club memberships, most golfers seem to have little in common with the unwashed eco-warrior brigade. The divide between the two groups is not just sartorial, but stems from the fact that many golf clubs use huge amounts of water, disfigure the landscape and use fertilisers and pesticides to keep their greens lush. However, this gulf may soon be bridged, as a Cambridgeshire club which boasts a full-time ecologist, not to mention a resident stoat at the eighth hole, is poised to become the UK's first organic golf club.

The owners of New Malton Golf Club, an 18-hole course, claim it has been chemical-free for a year, and they plan to apply to the Soil Association for organic certification. The course's out-of bounds areas are home to birds, including woodpeckers, kestrels, owls and pheasants, as well as hares. The owners plan to graze animals on the land, while also growing fruit and lavender.

"We don't use any pesticides and have been 100 per cent chemical-free for a year," said the golf course's co-owner Paul Stevenson. "We get water from the River Cam to water the greens... there will be traces of chemicals in it, but we are hoping to find a bore hole to get around that."

Prof. Golf Association - UK

In India, many new Golf Courses have started coming up and there is interest in ecofriendly inputs. Cricket pitches and cricket fields are another major user of agro chemicals. Some of the progressive clubs showed interest in adopting eco-friendly bio inputs, though many are conservative and sceptical. Field demonstrations and patient interactions become important. Before that of course, based on discussions and local conditions the package of bio inputs have to be selected. One important criteria here is that they have to be compatible with the conventional chemical inputs (though application things may be different) and the change would be gradual. Table : 1 is a typical application chart prepared by a dedicated groups of scientists of New Age Corporation, Gujarat Life Sciences and The Science Ashram teams in collaboration with the curators.

A typical package consisted of three types of products as given in Table 1 and Table 2.

Table 1: Organic Turf Management Inputs

		Soil Enrich (Gls Enrich)	**Multifunctioning Broad Spectrum Products** • All in one - GLS ENRICH - for soil enrichment and conditioning - made from Microbial Consortia Technology • Unique Functioning – SUPERBIOLIFE-for enhance capacity of the plant against various soil born pathogens and diseases. • Dual functioning - PRO-GROW - for root establishment and disease control • Multi functioning - GLS TRANSFORM - for leaf blade strength, preventing leaf blade damage, providing bright green colour
		Root Establi-shment Superbiolife & Progrow	
		Lush Green Lawn Gls Transform (Growth Enhancer)	

Table 2: Bio input package for Turf

Gls Enrich (Bio Compost)

For Nutrient Management

Formulated in Specialized Granular form for Slow release of Macro & Micronutrients.

Fortified with Microbial Cultures providing Nutrient and Disease Control.

- Helps to boost the Soil Fertility by Increasing the Soil Micro Flora.
- Enhances the Soil Structure and Porosity.
- Increases the Disease Resistance Power in Grasses.
- Helps to reduce application of High Chemical Doses.

Table 2: (A)

Progrow

For Root Development and Disease Control

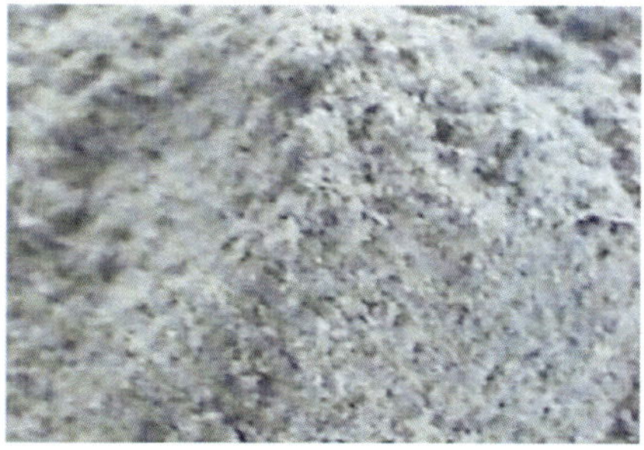

Combination of Various Fungal and Bacterial Microorganisms. Liquid Formulation is Enhanced With Various Natural Bio Agents.

- Creates a Formidable Micro-Ecosystem around the Roots, Enhancing their Growth and Strengthening them.
- Increases the Micronutrient Uptake from the Soil by the Grasses.
- Help to Control Fungal Attacks and Prevents drying off.
- Provides much required resistance under stress Conditions.

Table 2: (B)

Superbiolife

For Overcoming Stresses and Providing Green Luster

Unique Formulation of a Wide Range of Natural Extracts.

- Provides the Strength and resistance against the Biotic and Abiotic Stresses.
- Regulates and Enhances the growth of the Grass.
- Stimulates the Metabolism.

Helps to Provide Rigidity to the Grass Blade and restore Luster, Shining and Greenness of Grass.

Table 2: (C)

In developing the special packages for the turf, the ease of application and the prevailing practices were considered to enable adaptation of the new products easily. In the beginning, as many as 8 – 9 products were used each with specific functioning. This was not easily acceptable and demand to have only 3 or 4 products with multiple functions became clear. This needed special efforts and formulation developments, small farm trials and demonstrations. Table – 2 lists such products and their functions.

In Table – 3, the guidelines for field application is given. Depending on the duty of the turf the application amount and frequency have to vary e.g. in cricket field, the pitch area and the area near the wicket cricket and bowler needed higher amount and frequency of application than the outer fields. Similarly, for Golf courses, the experienced curator uses his judgement based on the duty on the turf. Photograph – 1 gives the focus area for the turf management.

Table 3: Application Process

Gls Enrich :

Broadcasting manually
Outfield : Once in a month
Pitch : Once in a week

Pro-Grow and Gls Transform:

Mix in 1:1, Dilution in water and spraying by Spray pump.
Outfield : Once in a month
Pitch : Once in a week

Baroda Cricket Association Ground

The results of these applications in different areas and situations are given in the photographs 1 to 3. (Photographs of Fields – Before and After Treatment). Photographs of results of treatment as discussed above for Cricket Grounds as well as Golf Courses show positive results and are valuable demonstrations that can be repeated in many such cricket fields and Golf Courses.

Regeneration of Grass in the Outfield

The luster and greenness of turf has improved significantly

(Before)

(After)

Photograph - 1

Regeneration of Grass in the Outfield

Enhancement in the soil micro flora leading to better fertility and healthy turf

Photograph - 2

Improvement of Fungal / Diseases Infected Areas

The grass established well in the infected area and has shown good root networking and growth

Photograph - 3

From the above extensive work, it became clear that the turfs for Golf Course, cricket field and School Grounds with the right combination of Eco-friendly bio inputs give the following benefits as per the feed back from the respective curators.

- Improved soil and air quality
- Healthy and sustainable growth
- Reduced watering
- Reduced costs

Though exact data are not yet available, in general it was reported that the amount of water required reduced substantially and the amount of chemical fertilizers by about 30% and the players were happy. The results seem to improve year after year and the possibility of eliminating the chemical inputs and having totally organic inputs is also visualized.

CONCLUSION

There is great concern for eco-friendly sports fields for the obvious reasons. However, there is a slow rate of acceptance and change over. Also the lack of awareness and resistance to change to alternative practice is a major reason. Ready availability of alternative eco-friendly input is another constraint. Demonstrable results should go a long way in acceptance of alternative eco-friendly practice.

REFERENCES

1. Organic Turf Management Field Report of New Age Eco Solutions and Gujarat Life Sciences, 2011.
2. Organic Golf, Newday, July 1, 2003.
3. Thankamani C.K. et al, Nursery mixtures, Journal of Spices and Aromatic Crops, Vol. 16(2), 2007.

Bin Pani Sab Soon ! (Without water everything is zero !)

— Rahim

14

More Crop Per Drop- Improving Water Use Efficiency in Agriculture for Sustainability

It is estimated that the country needs about 245mt of food grains, 30mt of sugar and jaggery, 7 lakh of tea, 1.7 lakh t of coffee, 10mt of vegetable oil and 17m bales of cotton. For realizing this, irrigated agriculture has to play an important role. To achieve this water use efficiency (WUE) in the irrigation sector has to be improved considerably as bulk of the available water is used in agriculture. But, with the ever-increasing population and industrialization problems, the demand for water by these sectors is bound to increase manyfolds in future. The requirement of water by different sectors during 2025 is estimated to be 105 m ham, which is 30 m ha m more than the present one. The share of water utilization for agriculture is expected to get reduced from the present percentage of 8~ to a value of 73 by 2025. Further, the demand for water for agricultural purposes is estimated to increase from 470 krrr' to 740 krrr' in a period of four decades (Table 1). During the same period there will be four-fold increase in the water demand for non-agriculture purposes.

Table 1: Water demand pattern in India (Km^3)

Area	1985	2025
For agriculture	470	740
For non agricultural use	70	280

The per capita water availability, in terms of average utilizable water resources in the country, which was 6008 m^3 in 1947 (presently 1250 m ') is expected to dwindle down to 760 m' by 20 50 (TERI, 2003). All these emphasize the need for water conservation and improvement in use efficiency, to achieve "More Crop per Drop".

IRRIGATION STATUS AT THE NATIONAL LEVEL

Out of the 140 m ha of net sown area in the country, 50 m ha ia under irrigation working out to a percentage of about 35. In the different states, the net irrigation per cent varies from 12 in Mizoram to as high as 90 in Punjab. Though at national level the contribution from canal and under ground water are almost equal, yet in Assam, Himachal Pradesh, Orissa and the other North eastern states, it is predominantly canal system, while in Gujarat , Punjab, Rajasthan and Uttar Pradesh and ground water predominates as the source. In Tamil Nadu, tank irrigation is an important irrigation source with almost Quarter of the irrigated area under this system. In the other southern states as well as in Orissa also tank system is in vogue to a considerable extent.

The unscientific water use in both canal and lift irrigated commands has resulted in raising water table in the first case and fast receding ground water table in the second case. Thus, there is an imminent need for reducing the irrigation water application in agriculture with out affecting the yield and there by increasing the Water Use Efficiency (WUE) The options for increasing the water productivity/improve the WUE are discussed here.

PREVENTION OF WATER LOSSES IN TRANSPORTATION SYSTEM

Huge quantity of water is lost during the transportation of water from the source (reservoir/main canal! tank/ wells) to the field and till it reaches the root zone. Hardly 20 to 40 per cent of the water released reaches the root zone. The irrigation efficiency can be increased by the selection of proper method of irrigation along with proper design. Apart from the high evaporative and seepage losses occurring at the reservoirs and main canal, considerable amount of water is lost as seepage and evaporation in minors, distributaries, field channel *etc.* The losses occurring in a typical project in the alluvial plains as reported by the Second Irrigation Commission is given in table 2.

Table 2: Losses of irrigation water(%) in unlined canal distribution system

Sources of losses	Seepage	Evaporation	Total
Main canals and branches	13.6	3.4	17.0
Distributaries	6.4	1.6	8.0
Field water courses	16.0	4.0	20.0
Field application losses	13.2	3.3	16.6
Total	49.2	12.3	61.5

In none of the projects in India, the total conveyance loss is less than 50 percent. In many cases, in the hotels etc. in the channels add to the losses through leakage.

IMPROVEMENT OF WUE IN PADDY

Paddy is the most water consuming crop and nearly 40 per cent of the total irrigation water available is used for paddy. Hence, any small saving in irrigation water in paddy will go a long way in the overall water available for other crops.

a) Reduction in percolation losses

One of the major sources through which the water losses occur in the paddy crop is through percolation. The studies conducted at the various AICRP centers on water management has shown that by adopting the improved puddling through power tiller along with cage wheels substantial percentage of water loss ranging from 10 to 25 in different locations of varying agro climatic and soil conditions can be avoided.

b) Optimum ponding depth

There exists a mis-conception in the minds of the farmers that, since paddy is high water requiring crop, the yield could be increased with the increase in the submergence depth. But the experiments conducted at different locations in the country showed that the crop needs only shallow depth of 5 cm and the higher depth does not have any benefit in terms of water productivity (Table-3).

Table 3: Effect of ponding depth of water on paddy yield and WUE

Locatin (state)	Ponding Depth (cm)	Grain Yield (t/ha)	Irrigation water requirements (cm)	Water Expense Efficiency kg/ha.cm
Kharagpur (West Bengal)	5101520	5.285.274.744.53	124183246297	42.528.819.3
Hyderabad (Andhra Pradesh)	510	4.594.42	167366	27.512.3
Chiplima (Odisha)	510	4.554.48	4962	92.872.2

c) Alternate wetting and drying in paddy

As rice is a semi aquatic plant, farmers feel that the crop need be kept always in submergence regime as most of them do presently. But, the results of the well laid out experiments through out India at the various Water Management Centers have clearly shown that the crop does not need continuous submergence, but actually requires/drainage varying from 1-3 days depending upon the soil type and climatic conditions.

Table 4: Effect of drainage on paddy yield (t/ha)

Location	State	Continuous Submergence	1 day drainage	3 day drainage	5 days drainage	Water saving (%)
.BhavaniSagar (T.N)	Tamil Nadu	4.57 (136)	5.09(110)	4.95(80)	4.16(55)	41
Bilaspur(M.P.)	Madhya Pradesh	5.86 (98)	5.59 (79)	5.13 (70)	4.62 (56)	29
Chalakudy (Kerala)	Kerala	3.47 (140)	3.44 (63)	3.45 (49)	-	65
Chiplima (Odisha)	Odisha	7.21 (99)	6.66 (90)	6.47 (76)	5.85 (66)	23
Hisar (Haryana)	Haryana	5.66 (220)	5.15 (196)	4.69 (126)	-	43
Kharagpur (W. Bengal)	West Bengal	6.11(197)	5.98 (150)	5.89(129)	4.99 (108)	34
Kota (Rajasthan)	Rajasthan	5.41(145)	5.33(86)	5.05 (68)	-	53
Ludhiana (Punjab)	Punjab	5.52(190)	5.44(145)	5.12(113)	5.20(96)	40
Madurai (T.N.)	Tamil Nadu	6.59(128)	6.66(104)	6.18(82)	-	36

By following this alternate wetting and drying technique, 25 to 60 percent of water input to this crop can be reduced in the different soils and agro climatic conditions of the country. Most of the centers have demonstrated this technology on the farmers' fields on a large scale successfully. In UkaiKakrapar command of South Gujarat, by adopting this technology, about 15 per cent more area was brought under irrigation in some minor with the same quantity of available water. In China also this technology is widely adopted.

d) Irrigation techniques for arable crops other than paddy

Irrigation water is applied to the crops through different methods *viz.* surface, sub surface pressurized systems *etc.,* of which surface method is the oldest and most common one. In this method, the water is 'applied directly to the soil surface from a channel or from an open ditch at the top end of the land. Normally, surface irrigation methods are poor in their irrigation efficiencies, but the different technical options available can increase them. Lining of the conveyance system, land leveling, adoption of appropriate irrigation schedule for the different crops as per the water table conditions, selection of suitable and viable cropping sequences under different water availability conditions, improvement in the existing designs of the surface irrigation systems *etc.* are some of the ways to increase the input efficiency of water for agriculture.

In the heavy black soils of Navsariin South Gujarat, the normal sugarcane yield could be obtained even with 55 per cent of the normal quantity of water, when applied in alternate furrow along with sugarcane trash as mulch. Similarly, in Bhatinda the alternate furrow system could reduce the irrigation water for maize by 50 per cent. Under Pusa conditions, the irrigation water requirement of maize can be reduced by one third with the adoption of alternate furrow irrigation. Working on the same principle, the agro technique of paired row planting which can reduce the area to be irrigated can reduce the irrigation water considerably. Water savings to the tune of 70 per cent were reported from Hissar due to adoption of paired row system in close grown crops like *bajra* and *wheat*.

e) Adoption of mulching

Mulching is the practice of spreading any extraneous material on surface soil to increase moisture conservation, reduce evaporation, moderate soil temperature, check weed growth and some times control soil borne pests and diseases when they are used as solarising agents. The improved edaphic environment created by mulching enhances the water and nutrient use efficiencies. The concept of mulching is not new to Indian agriculture as it is being followed in rainfed agriculture as intercultivation, hoeing *etc.* which is nothing but soil mulching.

Of late, it is being adopted in irrigated conditions also. In both situations either organic or plastic *mulches* have proved effective in reducing the irrigation requirement as well as to conserve the moisture. Many trials conducted in different crops in the Gujarat state have shown that the use of plastic mulch can reduce the irrigation requirement by about 50 per cent and that of organic mulches by about 30 to 35 per cent in different crops.

Simultaneously, there has been considerable increase in the yield of different crops. Paddy straw or wheat straw has been reported to be good mulch material by different workers. But, since these materials are fed to the cattle as cattle feed their scope as mulch material is limited. Hence, farm wastes like sugarcane trash, castor shell and similar farm wastes, which have no other economical value should be promoted as organic mulch materials. The field and aquatic weeds, have been effectively utilized as mulch materials. In the high rainfall hilly areas of Palampur, use of the field weed Lantena as mulch material enhanced the yields of *kharif* maize and *rabi* wheat (Anon. 1996). Utilization of the aquatic weed, water hyacinth increased the yield by almost 50 per cent and there by the moisture use efficiency (Anon., 2001-02). In the South Gujarat region, use of field weeds as mulch increased the yields of cotton and pigeon pea by as high as 70 per cent. Though, plastic mulch is technically the most effective, yet with the present cost of the film it is economically viable only for cash crops and high value vegetable crops and it is economically viable only for cash crop and high value vegetable crops and it is the best mulch material in salt affected soils and when saline water is used for the irrigation (Table-5) A combination of mulching in green houses give very high water use efficiency.

Mulching inside Green House – High WUE

Table 5: Mulching technology for water savings and yield improvement

Crops	Location (State)	Water saving (%)	Yield increase (%)	Remarks
Amaranthus	Tavanur (Kera)	-	84	-
Banana	Navsari (Gujarat)	33-50	13	Sugarcane Trash mulch/ Black Plastic
Ber	Sardar Krushi Nagar (Gujarat)	N.A.	25	Rainfed / Black plastic
Brinjal	Achalia (S.K. Nagar& Kandha (Gujarat)	-	10	Black plastic
Cabbage	Jorhat (Assam)	50	10	Black plastic
Castor	Sardar Krushi Nagar (Gujarat)	N.A.	75	Rainfed/ Castor shell/ Mustard straw
Cauliflower	Hisar (Haryana)	-	33	
Pigeon pea	Navsari (Gujarat)	-	47	Sugarcane trash
Sugarcane	Navsari (Gujarat)	28	10	Sugarcane trash
Tomato	Sugarcane trash	44	53	Black plastic

f) Adopt precision methods of irrigation

Advanced irrigation systems like drip, sprinkler, mini sprinkler, bubbler, micro tube *etc.* are some of the techno-economically tested and viable innovated options for using less water, improve the yield and there by the WUE. These pressurized irrigation systems are especially useful in shallow soils, soils with steep slope andlor undulating topography, where the land leveling concept may not be a viable proposition, highly permeable soils, in the cases of application of low depth of water (less than 5 cm) *etc.* Apart from the above, the use of pressurized irrigation system can result in saving of 5 to 8 per cent of the land which is lost otherwise for the formation of channels, bunds *etc.* Even moderately poor quality water can also be used with this technique. These systems can be termed as precision irrigation systems since the exact quantity of water needed is applied at the right time. While the sprinkler system is recommended for close grown and row crops, the drip and other micro irrigation systems are recommended for horticultural crops including orchards, vegetables, widely spaced flower crops like roses *etc.* and agricultural cash crops like cotton, sugarcane *etc.* Presently only 3 lakhs hectares are under drip irrigated.

g) Micro irrigation systems

Micro irrigation is an irrigation system with high frequency application of water in and around the root zone of the plant system. It essentially consists of any emitter device like drip, mini sprinkler, micro sprinkler *etc.* along with laterals, main line, control head system with the fittings.

Though, the present day drip system is new to India, but conceptually the indigenous micro irrigation has been invogue in the farm of bamboo pipes, perforated clay pipes, pitcherporous cups *etc.* since long. While the bamboo system is common in the Meghalaya for beetle vine, pepper and arecanut grown on hill slopes, in Rajasthan the pitcher! porous cups irrigation has been in practice mainly for the vegetable crops. Though, as against the potential area of 27 m ha, presently only 3 lakhs hectares are under drip, yet the growth rate is highest in the world with a figure of 196 per cent during the period 1993-94 and 1997-98. In the late ninties through GSFC Science foundation, a company Gujarat Agri Processing Ltd. As the founder M.D. of the company, a 10 year vision plan was made and later the company was renamed as Gujarat Green Revolution Co. (GGRC). The major emphasis was on promotive drip irrigation systems with proper linkage between farmers and suppliers of the drips. The successful growth of this company and use by farmers has set an excellent example.

Under the NCPA, New Delhi, and AICRPs on water management. Dry farming and different horticultural crops of ICAR, New Delhi, these systems of economically viable location are being evaluated for various agro climatic regions and their mandate crops.

The trials conducted in different parts of the country have shown that 20 to 70 percent water application to the crop can be reduced for different crops of various agro climatic regions by adopting any one of the micro irrigation technologies, Yield increase to the tune of as high as 60 percent has been observed (Table: 6)

Micro Irrigation for Small Nursery

Table 6: Response of crops to drip irrigation

Crops	Water saving (%)	Yield increase (%)
Ash gourd	12	10
Banana	35-45	5-30
Brinjal	26-65	16-63
Castor	25-70	30-36
Cauliflower	12-44	28-72
Cotton	40-50	5-35
Papaya	54-68	19-77
Potato	0-54	25-66
Sugarcane	14-41	16-40
Turmeric	53	76
Watermelon	36	88

The simultaneous changes in the water saving and the crop productivity affect the water productivity. Sivnappan analysed certain typical cases for different crops and found that while in potato where no water saving was observed, the yield increase was to the tune of 46 per cent with a water productivity figure of 46. On the other hand, in cotton, where there was 60 per cent water saving with a simultaneous increase in the yield to the tune of 25 per cent, the change in water productivity was maximum with a positive figure of 255. Thus, adoption of micro irrigation will be specially beneficial in the areas of water scarcity (Table: 7)

Table 7: Water productivity changes due to drip

Crop	Change in the yield (%)	Change in the water use (%)	Change in water productivity (%)
Banana	+52	-45	+173
Cabbage	+2	-60	+150
Cotton	+25	-60	+255
Grapes	+23	-48	+134
Potato	+46	0	+46
Sugarcane	+33	-65	+205
Sweet potato	+39	-60	+243
Tomato	+50	-39	+145

In the tribal and hilly areas with very small land holding and unreliable electricity supply, low cost drip systems can help the farmers in a great way. Insuch type of areas, the low cost drip systems like bucket drip, drum drip, pitcher irrigation *etc* are becoming more popular.

h) Drip and mulch combination for furthering improvement of water expense efficiency

A combination of the drip technology along with mulching has been reported to have a synergistic effect in terms of not only water saving, but also the improvement in yield (Table 8). This technology almost doubled the yield over drip alone, in many crops with special reference to vegetable crops like tomato, okra, chillies, capsicum *etc.* will an interesting study in Gujarat while drip alone could not influence the yield of brinjal, the super imposition of mulch technology with drip increased the yield by 41 per cent over drip alone. The net result of this combination was improving the ultimate water expense efficiencies in different crops by as high as 200 to 460 per cent.

Table 8: Impact of drip with mulch on the water utilization and yield of different crops.

Location	State	Crop	Yield increase (%)		WUE increase (%)	
			Drip	Drip Mulch	Drip	Drip Mulch
Kharagpur	W. Bengal	Tomato	26	57	110	162
Navsari	Gujarat	Tomato	31	56	106	179
Khandha	Gujarat	Tomato	6	31	145	204
Pantnagar	Uttarakhand	Okra	4	24	56	87
Anand	Gujarat	Chillies	50	70	65	87
Solan	H.P.	Pea	50	75	98	131
Solan	H.P.	Apricot	22	43	51	120
Solan	H.P.	Peach	65	75	106	118

Ref : NCPA, New Delhi, 1999

If we could convert a sizeable part of the irrigated areas into precision irrigation systems, there will be considerably more area that can be brought under irrigation along with increasing the land and water productivities. The potential area those can be brought under drip and sprinkler in India is 27 and 39.5 m ha, respectively ie. a total of 66.5 m ha (Table 9).This is as against the present total area of only 0.5m ha.

Table 9: Theoretically potential area under drip and sprinkler in India (m. ha)

Crop	Area (m ha)		
	Drip	Sprinkler	Total
Cereals	-	27.6	27.6
Pulses	-	4.2	4.2
Oil seeds	1.9	1.1	3.0
Cotton	9.0	2.6	11.6
Vegetables	5.3	2.0	7.3
Spices and condiments	1.4	1.0	2.4
Flowers and medicinal and aromatic plants	-	1.0	1.0
Sugarcane	4.1	-	4.1
Fruits	3.4	-	3.4
Coconut	1.9	-	1.9

i) Green house cultivation

This technology was introduced in the country as a part of the package of "Israeli Technology". But the green house cultivation as adopted in Israel is not applicable into in our farmers' conditions as they are high energy consuming ones, requiring lot of electrical power in uninterrupted condition" Hence, a naturally ventilated green house was developed and tested different agroclimatic conditions of the country. This type of green house is specially suited in tribal and hilly areas where water availabilities is a constraint and farmers are economically poor with small land holdings. Studies conducted at different locations have proved the advantage of green house cultivation in considerably increasing the yield and the WUE (Table 10). Low cost green houses could be quickly erected after the earth quake in Kutch - Gujarat in 2000 to provide temporary shelters. Subsequently, they became a part of livelihood programme for growing leafy vegetables like spinach, coriander, chillies, tomato lucerne etc. and nursery for horticulture. (Photo)

Table 10: Green house technology for improving WUE

Crop	Location	State	Yield (t/ha)		WUE (kg/ha.mm)		Improvement in WUE (%)
			lakhs	lakhs	(flower)	(flower)	
Chrysanthemum	Bangalore	Karnataka	20.2	10.2	48	27	78
Tomato	Navsari	Gujarat	108	48	345	153	125
Rose	Navsari	Gujarat	1.5	1.0	118	78	51

Micro Irrigation in Low Cost Green House - Kutch

Improvement of WUE in the canal command areas

The most important aspect affecting presently the WOE in crops in the canal command areas is water logging and as a result of which the drainage system problems. In our country water logging in the canal command areas are increasing at an alarming rate. In the Tungabhadra project of Karnataka more than 33,000 ha are water logged. Similar grave situations occur in most of the commands in the country. As per a report of the National Commission on Agriculture (1976) the water logged area in some major states in the country was totalling to a figure of 6.0 lakh hectares (Table 11). Subsequently, the Ministry of Agriculture has estimated it to be 8.5 lakhs ha, though, it has to be still field verified. About 60 per cent of the water logging is surface water stagnation during *kharif* and rest is due to the water table fluctuations affecting crop growth.

State	Area
Andhra Pradesh	3.39
Bihar	1.17
Gujarat	4.84
Haryana	6.20
Punjab	10.9
Rajasthan	3.48
Uttar Pradesh	8.10
West Bengal	1.85

GAU initiative for Narmada canal water

Progressive work was being taken up in a big way. Through our teams of experts at GAU(Gujarat Agriculture University). I had taken up detailed discussions for future strategy. It was estimated that once the network of Narmada canal is completed, the irrigation potential of Gujarat state will double. We also discussed the need to see problems along with the new opportunity. Experience in South Gujarat (after Ukaikakrapar), Rajasthan (after Indira Gandhi Canal Projects) of over use of water, water logging and giving lower farm productivity was alarming. The farmers of water scarcity areas (like North Gujarat, Saurashtra) were expected to be over enthusiastic about the "new water" and a model of judicial was utmost necessary. We planned and implemented 6 demonstration units at different locations around the canal. Demonstration of proper use of water for different crops and under different agroclimatic conditions because a major mission.

Rise in the water table conditions, creating waterlogged situation in root zone of the crop is a very common phenomenon in most of the canal command areas in India. Under these conditions considerable amount of ground water may enter the root zone through capillary action and there by meet part of the ET demand of the crop. This contribution will depend upon the soil type, water table depth, rooting depth of the crop prevailing weather conditions *etc.* Thus, under shallow water table conditions, there is a definite possibility for reducing the irrigation water to be applied for the crop. Though, the irrigation water requirements of different mandate crops of the various agro climatic conditions of the nation have been worked out, yet they are for the normal water table conditions and the irrigation requirement will vary depending upon the water table and the rooting pattern of the crop.

The waterlogged conditions tend to reduce the crop yield. In Ukai Kakrapar command in South Gujarat, the high water table conditions (water table fluctuating from surface to 1.5 m) are reported to reduce the yield of sugarcane by 30 per cent. The yield loss could be retrieved with proper drainage. An operational trial on drainage in sugarcane on the farmers' fields in the Ukai Kakrapar command has shown that the surface drainage provision increased the yield of not only sugarcane but also the yields of paddy, gram and Indian bean (Table 12).

Table 12: Impact of Sub Surface Drainage on Yield (t/ha) and WUE improvement in different crops

Crop	Yield		Improvement in WUE (%)
	Before drainage	After drainage	
Sugarcane	68	92	42
Paddy	3.4	4.0	20
Gram	0.83	1.1	33
Indian bean	0.50	0.75	50

Research on agricultural drainage was carried out in several ICAR institutes and State Agricultural Universities, both on surface and sub surface drainage. Recently, the trials conducted under the Indo- Dutch Project on Drainage at Karnal (Haryana), Navsari (Gujarat), Bapatla (Andhra Pradesh), Bheemarayanagudi (Karnataka) and Hanumangarh (Rajasthan) have established the techno economic feasibility of close sub surface drainage, with the use of perforated PVC pipes, for the crop production and improving WUE in water logged saline soils (Reference Photograph).

Subsurface Drainage to improve Sugarcane productivity

Poor quality water and effluent usage

All waters including good canal water do contain salts. The quality of river water is mostly good. But, the waters of Chambel, Tapti, Yamuna, Godavari and Krishna are slightly saline. More than 32 per cent of ground water surveyed showed that they are either saline or sodic or both. Ground water surveys indicated that poor quality waters are used as irrigation source in 1.5 m ha in India (Reddy and Reddy, 2000). The categorization of water as to its salinity hazards depends upon the soil texture, drainage conditions and crop to be grown and not on the *perse* conductivity value. While moderate saline water cannot be used in heavy soils with poor drainage, saline water can be used in a sandy soil with no limitation of drainage. Barley, sugarbeet, cotton, wheat, safflower, soybean, sorghum, pearlmillet, rice and mustard are some of the agricultural crops, which can be irrigated with reasonably saline water (5-10 dS/m). Agro techniques with the planting method, consideration of the leaching requirement, irrigating as per the sensitive stages of the crop *etc.* will improve the WUE with saline water. When waters of different qualities are available mixing both or cyclic use of them will not only improve the yield and WUE but also sustain the soil health. A study conducted at Ludhiana showed that the yield of wheat-and rice can be 'sustained by cyclic use of canal and saline water (Table 13).

Table 13: Effect of saline water use through drip with mulch in Brinjal

Treatments	Fruit Yield (t/ha) (3 years pooled)
Variety Surtiravaiya	
Dark purple round	35.11
CD at 5%	31.43
Mulch	1.81
No Mulch	30.49
Black poly mulch	36.85
CD at 5%	1.81
Salinity levels (Ds/m)	
Control (Normal water)	34.71
4	40.06
8	32.56
12	27.30
CD at 5%	2.56

In the normal conditions and with surface method of irrigation even 2 of 4 ds/m water cannot be used in these soils. The sodic water also can be profitably used provided enough care is taken to provide the drainage and gypsum added to the soil as per the gypsum requirement of the water or the water should be passed through the gypsum tanks blocks. With the ever-increasing industrialization and urbanization, huge quantities of effluents are generated which pose a disposal problem. Though, quantitatively these can be foreseen as good alternate source of water resource, yet their suitability depends upon majorly the organic and inorganic pollutants they contain. Not all the industrial effluents are useless. Reports exists on the usage of paper mill effluent and spent wash from alcohol industry as potential sources of alternate water resources. But, they have to be viewed with sufficient care and potential pollution hazard should be looked into. In Gujarat, the effluent from acetone factory has been found to be devoid of any pollutants and technology has been developed for the use of the same as source of irrigation water. When mixed with the canal water in a dilution of 50 times, the effluent water use not only improved the sugar cane yield, but also reduced the nitrogen requirement by 50 per cent and there was no need for application of P and K. The subsequent soil and plant analysis also showed no accumulation of any heavy metal.

The WUE can be enhanced only by adopting the whole package of practices for the crop like, optimum plant population, planting time, spacing, fertilizers, weed management, pest management *etc.* Highest yield and WUE is obtained only in optimum combinations of all the inputs and not individually. For example, adopting drip technology in isolation will not result in the maximization. It has to be along with proper crop geometry, fertigation *etc.* The trials conducted by the different Plasticulture Centers of the National Committee on Use of Plastics in

Agriculture has shown that with fertigation along with drip, not only ensures the fertilizer savings, but also improves the WUE, thus enhancing the effectiveness of the drip. Water/moisture and fertilizer interaction is a very important one in irrigated as well as dry farming situation. Inthe absence of optimum fertilization the response to irrigation will be reduced and *visa versa.* Similarly, in dry land agriculture, fertilizer quantity and time of application is much dependant on the soil moisture. Application of organic manures will enhance the WUE by providing better water availability to the plants irrespective of the soil texture, both in irrigated as well as rainfed condition.

In Tigray area of North Ethiopia, earlier facing the worst drought for a generation, use of organic compost in secured villages improved the productivity of Maize, Millet, Barley and pulses etc substantially and with much lower consumption of water. A remarkable improvement was reported in the sustainability and livelihood of the region by adopting a holistic approach.

CONCLUSIONS

1. Though the government of India has invested huge amounts for the development of irrigation facilities during the different Five Year Plans, yet the efficiency achieved in irrigated agriculture is far below the desired levels. Apart from the need of improvement in the water distribution system to improve the conveyance and irrigation efficiency, there is utmost need of proper adoption of the water management technologies developed at the various State Agricultural Universities (SAUs) of India to improve the water use efficiencies to have more crop per drop. Systematic canal operation has been found to improve the irrigated area for the same quantity of water by 18 to more than 100 per cent in different canal commands of the country. Among the different aspects, selection of proper irrigation method plays a vital role.

2. The R&D efforts to popularize the micro irrigation technologies has resulted in the increase in area under micro irrigation several folds. Crops grown under the cost effective low cost effective, low cost green houses can increase the yield of crops like tomato, bhindi, lucerna, roses by two to three folds with almost 50 percent less water. Apart from fertigation, reducing the application of inorganic fertilizer to the crop in integrated way with the organic sources composts and bio fertilizer go a long way in improving the WUE.

3. Along with appropriate water management, timely application of all other inputs should be done to maximize the productivity, improve the Water Use Efficiency and finally reap “more crop per drop”

ACKNOWLEDGEMENT

Much of the data and field work was done by our GAU, Navsari team of the water management projects led by Dr. S. Raman, Dr. R.P.S. Ahlawat and Dr. K. Patil. These are well documented in Water Management Centre Reports between 1997 to 2007.

REFERENCES

1. Water Management – Reports of Navsari. Water and soil centre- Gujarat Agriculture Universities – 1997 to 2007.
2. S. Raman, R.P.S Ahlauwat and Mehta M.H., More Crop Per Drop – Improving water use Efficiency in Agriculture, GAU Annual Report – 2002.
3. Personal Interactions- Farmers Cooperative Groups – Narmada- Command Area – Central and North Gujarat.
4. Dalwai Ashok, Climate Change Adaptations- Agarian Perspectives, Global Water Meet Dhanbad – 24-26 Oct '2016.
5. Ismail Serageldin, "World commission Water for the 21st Century – Emerging Challenges" – USSRF, India Feb 3-5 1999.
6. Developments, Issue 21, 2003.

Two things are unlimited; the number of generations we feel responsible for and our inventiveness.

— Jan Timbergen
Nobel Prize - Economics

15

Agricultural Rehabilitation of Earthquake Affected People

The recent earthquake in Nepal brings back memories of 26th January 2001. While the nation was busy in celebrating Republic Day, an earthquake shattered Gujarat especially Kutch and Saurashtra regions. A large number of buildings and houses collapsed within moments. Thousands were injured and many lost their lives and many more became homeless.

Kutch received quick responses from all over the world and a large number of individuals, institutions, government and semi-government bodies, professional experts, religious and social organizations etc. sprang into action to bring back some normalcy in the lives. However, long-term rehabilitation program based on scientific planning is most vital after such disasters. Since agriculture was the main source of livelihood of majority of Kutch people, a plan based on

Picture 1
Bhachau – Kutch soon after the earthquake in 2001

scientific agriculture and integration of the same with livelihood, nutritional and long term social considerations was required. It has also been observed that new and appropriate technology can be fast adopted even by small and marginal farmer women and men under such circumstances.

GREEN HOUSE – CUM – SHELTERS

As the Vice Chancellor of Gujarat Agricultural University, I had organized meetings with our Scientists, NGOs and farmers. On checking with different locations in Kutch we realized that Shelter was the biggest problem. They were needed in very large numbers and immediately. At various campuses of GAU (erstwhile Gujarat Agricultural University), and the model farm at Baroda, different kinds of Green House Models were set up for experiments and training. It therefore occurred that low cost Green Houses (LCGH) can come to the rescue as immediate temporary shelters of people. These Low Cost Green Houses (LCGH) could be erected very quickly and were not expensive and required only local materials and would be easily erected by unskilled people.

A plan was prepared to set up such Shelter Green Houses. When Dr. M.S. Swaminathan, telephonically enquired about our well beings from Chennai – I assured him that we were safe and briefly mentioned about our plan. He encouraged me to send him a small write up which we did in the later hours of the same night. On the very next day we got a call from Dorabji Tata Trust, Mumbai, saying that Prof. Swaminathan has discussed the proposal in the trust meeting and a large support fund is sanctioned for this project. This was a great moral booster and immediately teams of students, staff, farmers etc were organized to set up the low cost green houses (LCGH) in different villages in Kutch.

Picture 2

Picture 3

Model LCGHs for immediate shelters

IMPLEMENTATION – CUTTING THROUGH PROCEDURES

In a short span of less than one month, nearly 1200 low cost Green Houses were erected with the help of Students and Staff of the University, NGOs and local farmers. These were not permanent structures, but provided great temporary relief to the displaced people in the biting cold in the Kutch as families could easily move in inside these temporary shelters.

BOX - 1

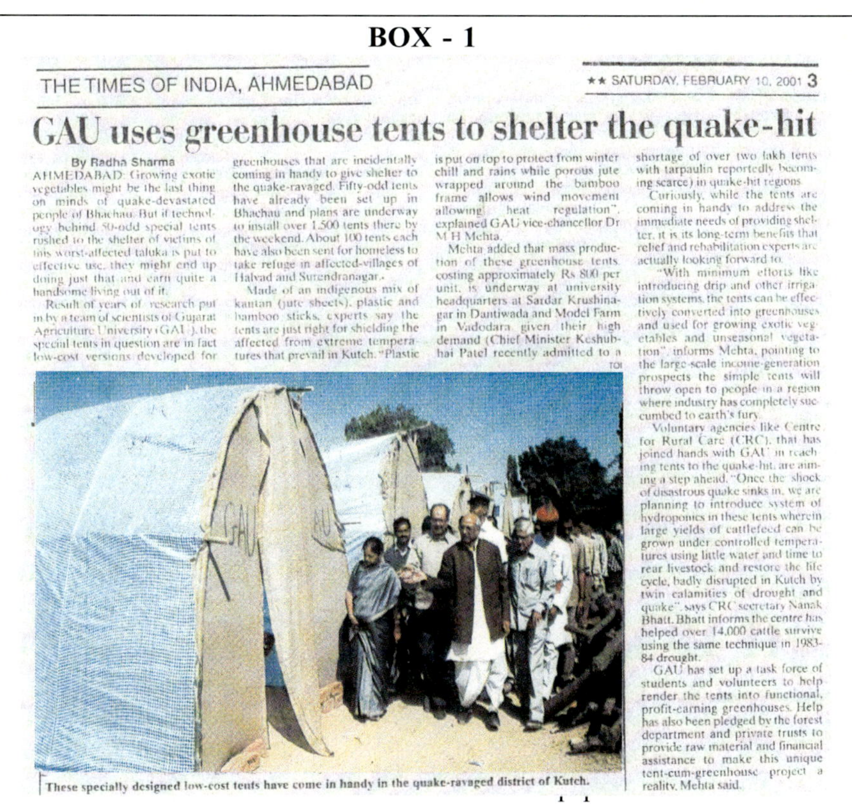

THE TIMES OF INDIA, AHMEDABAD ★★ SATURDAY, FEBRUARY 10, 2001 3

GAU uses greenhouse tents to shelter the quake-hit

By Radha Sharma

AHMEDABAD: Growing exotic vegetables might be the last thing on minds of quake-devastated people of Bhachau. But if technology behind 50-odd special tents rushed to the shelter of victims of this worst-affected taluka is put to effective use, they might end up doing just that and earn quite a handsome living out of it.

Result of years of research put in by a team of scientists of Gujarat Agriculture University (GAU), the special tents in question are in fact low-cost versions developed for greenhouses that are incidentally coming in handy to give shelter to the quake-ravaged. Fifty-odd tents have already been set up in Bhachau and plans are underway to install over 1,500 tents there by the weekend. About 100 tents each have also been sent for homeless to take refuge in affected-villages of Halvad and Surendranagar.

Made of an indigenous mix of kantan (jute sheets), plastic and bamboo sticks, experts say the tents are just right for shielding the affected from extreme temperatures that prevail in Kutch. "Plastic is put on top to protect from winter chill and rains while porous jute wrapped around the bamboo frame allows wind movement allowing heat regulation", explained GAU vice-chancellor Dr M H Mehta.

Mehta added that mass production of these greenhouse tents, costing approximately Rs 800 per unit, is underway at university headquarters at Sardar Krushinagar in Dantiwada and Model Farm in Vadodara given their high demand (Chief Minister Keshubhai Patel recently admitted to a shortage of over two lakh tents with tarpaulin reportedly becoming scarce) in quake-hit regions.

Curiously, while the tents are coming in handy to address the immediate needs of providing shelter, it is its long-term benefits that relief and rehabilitation experts are actually looking forward to.

"With minimum efforts like introducing drip and other irrigation systems, the tents can be effectively converted into greenhouses and used for growing exotic vegetables and unseasonal vegetation", informs Mehta, pointing to the large-scale income-generation prospects the simple tents will throw open to people in a region where industry has completely succumbed to earth's fury.

Voluntary agencies like Centre for Rural Care (CRC), that has joined hands with GAU in reaching tents to the quake-hit, are aiming a step ahead. "Once the shock of disastrous quake sinks in, we are planning to introduce system of hydroponics in these tents wherein large yields of cattlefeed can be grown under controlled temperatures using little water and time to rear livestock and restore the life cycle, badly disrupted in Kutch by twin calamities of drought and quake", says CRC secretary Nanak Bhatt. Bhatt informs the centre has helped over 14,000 cattle survive using the same technique in 1983-84 drought.

GAU has set up a task force of students and volunteers to help render the tents into functional, profit-earning greenhouses. Help has also been pledged by the forest department and private trusts to provide raw material and financial assistance to make this unique tent-cum-greenhouse project a reality, Mehta said.

TOI

These specially designed low-cost tents have come in handy in the quake-ravaged district of Kutch.

We realized that implementation of such programs need a different and unconventional approach. Co-ordination with various agencies, local population and volunteers is a challenging job. Combining compassion and efficiency with 'Human First' approach is the spirit that has to prevail.

FROM SHELTER TO LIVELIHOOD – THE NEXT STEP

After a few months, when situation improved somewhat, a meeting was organized under the leadership of Dr. Swaminathan at IARI Delhi to discuss about converting the Low Cost Green Houses to source of food. The Green Houses were converted for Livelihood program. The local farmers started growing Crops such as Spinach (Palak) , Tomato Corriander, Cucumber, Fodder etc. for the first time. In this water scarcity and saline region many of these vegetable crops were new to the farmers. Earlier attempts by our Agricultural Scientists to promote the new technology of Green House, vegetable growing was not accepted in spite of several attempts. Thus there was a quantum jump in adaptation of technology due to this crisis and farmers often came out with innovative improvisions.

Picture 4: "I entirely agree with you that the Eco Agri Model is the most suited for small farmers. This is also the aim of our bio-village paradigm of sustainable rural livelihood and ecological security." ***Dr. M.S. Swaminathan***

Picture 5: LCGH to grow vegetables

WAY TO ECO AGRI REVOLUTION

The above experiments got support from several individuals and organizations including Dorabji Tata Trust, Swaminathan Foundation and other organizations. (Entire plan has been recorded in a booklet "Agricultural Rehabilitation – Converting a Problem into an Opportunity" by M.H. Mehta and M.S. Swaminathan - GAU Publication – 2001). What was more interesting was that the local farmers adopted the 'New' technology quickly and comprehensively.

INTEGRATED FARMS

A second phase plan for integrated farming i.e Agri – Horti – Livestock was taken up alongwith eco-friendly agri bio inputs which emerged as the optimum model for small and marginal farmers.

Kutch is a semi-arid region. Most of the farms are small and the evapotranspiration is higher than the replenishment. However, there are several unique crops of the region. While preparing the rehabilitation program, the most sustainable and appropriate model evolved which is schematically represented in Fig. 1.

The third phase plan involving tissue cultured date palms, saline water agriculture was drawn out. All in all a major opportunity to start and implement an Eco Agri Revolution could be created.

LESSONS LEARNED

The above lessons, though in very different conditions could be of use in the agricultural rehabilitation of other areas also. The following additional points are important for planning a rehabilitation program.

i) Temporary and permanent shelters have to be designed and implemented quickly. These become extremely important for the affected families immediately.

ii) Subsequently, these become very good sources for livelihood of people to grow vegetables, fodder, agroforestry planting material etc.

iii) In most situations like this, where people affected are from the rural areas, agricultural should be recognized as the most important requirement for rehabilitation.

iv) Agricultural activities can be planned and implemented to give the best empowerment and self-sufficiency to people.

v) The agricultural rehabilitation has to be worked out by deciding about the local situation crop patterns, availability of seeds and other agro inputs.

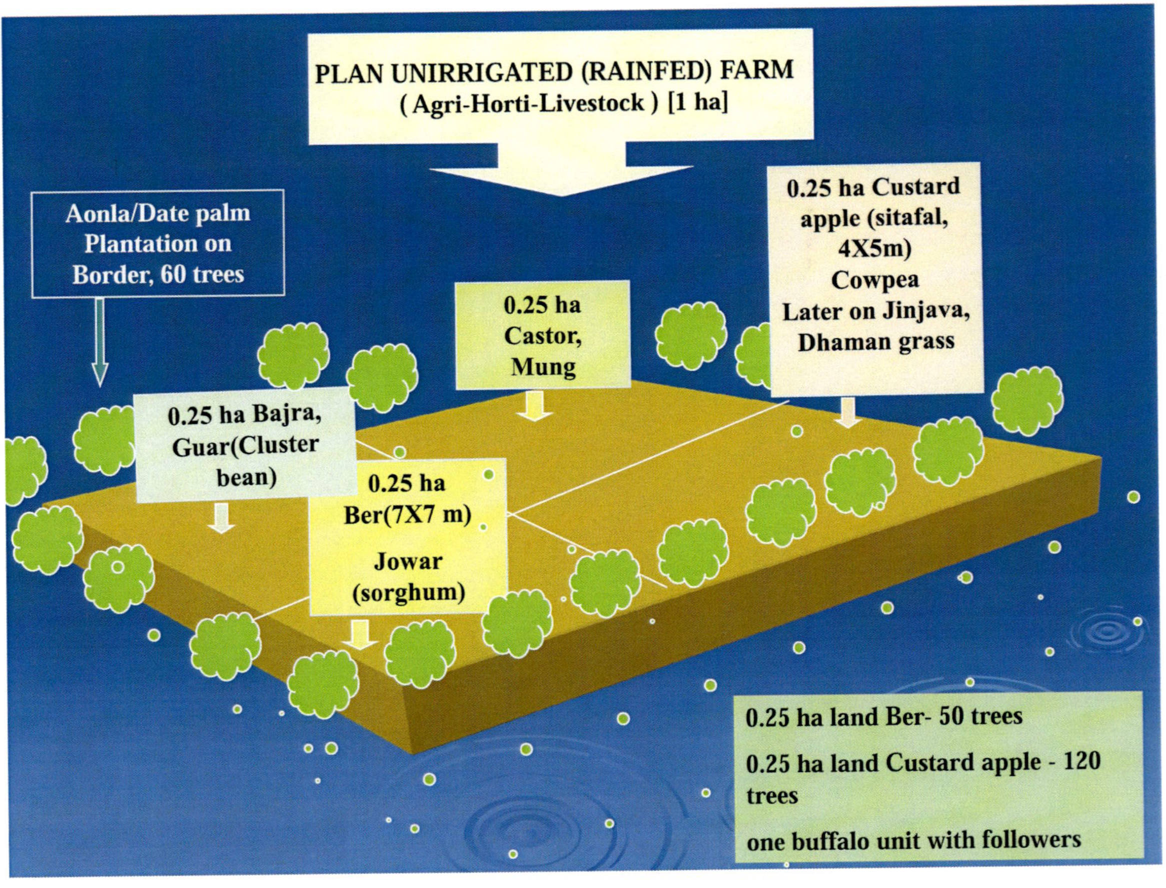

Fig. 1 : Integrated farm

vi) Massive immunization drive for the Livestock to be taken up.

vii) A comprehensive program including Biocontrol agents, low cost green houses, mixed cropping system can be drawn out and given as a package to groups of farmers.

viii) A complete package of Eco-Agri / Organic Farming can be adopted for such units. Biocontrol, Bio Composts and Bio Fertilizers packages can be adopted quickly and the extra produce can be marketed under a common brand (organic) for marketing.

ix) As given in the Figure, an integrated farming model – involving Agri – Horti – Livestock emerges as the most optimum model for small / marginal farmers.

x) It may also be viewed that a catastrophe like this can also give an opportunity to come out of inertia, generate new ideas and fresh thinking and planning and introduction of new technology on a large scale.

Picture 6: Bio Control Lab – Bhachau- Kutch

BOX – 2

The low cost green house revolution will help to enhance farmers' income and minimize risks arising from high temperature and low precipitation. Following the severe earthquake in the Kutch region of Gujarat, where also evapo-transpiration exceeds precipitation in most months of the year, the Gujarat Agricultural University launched a low cost green house program with financial support from the Sir Dorabji Tata Trust. The cost of a 200 sq. ft. low cost green house is about Rs.4000/- This has helped to open new vistas in water-saving agriculture. The expertise and material needed for farmers to take to such ecologically and economically advantageous systems of farming are available in the country.

Ref :SwaminathanM.S, in the book : "From Rio de Janeiro to Johannesburg".

Post Kutch experience as well as with our work in different areas / projects such as , Bioshield (Post Tsunami in Kalpakam – Tamil Nadu), and Uttarkashi (Post Landslides) – it is worthwhile to add some new and useful developments like:

a) Low cost Drip Irrigation System

b) Sustainable Integrated Agriculture model

c) Eco Agriculture i.e. low cost Agri Bio Inputs – The 20 : 20 model of higher Farm Production with lower input costs.

d) Agro Waste / Farm Waste used through Bio Composts using Multi-Microbial Technology.

e) Seed Ball Technology i.e. palleting the seeds with Agri Bio Products to different patterns of hilly areas to prevent land slide, greening, etc.

f) Co-operative Agro Processing Models

APPLYING THE KUTCH MODEL FOR NEPAL

It is worthwhile to consider this 'Kutch Model' to the situation in Nepal. In fact, it may be easier to take this up in Nepal where some more familiarity already exists for Green House Technology. The Kutch experience can be very handy to take up the phasewiseprogramof rehabilitation. The designs of model LCGH may need some changes / improvisations to suit local conditions.

BOX – 3

Agriculture Rehabilitation India can lend the Kutch Model to Nepal to recuperate their Agriculture

The earthquake that shook Nepal consumed the life of many people and has changed the landscape of the region forever. More than 8,800 people lost their lives and more than 23,000 were injured. The tremors that shook the region were the ones which left hundreds of thousands of people homeless reducing many villages into sheer memories. Once Nepal's pride, centuries-old buildings were destroyed at UNESCO World Heritage sites. The days to come will be difficult for Nepal as they have to draw up plans to pull this hilly state back to normalcy. Plans have to be drawn up to restore the lives of those who are left behind. Rehabilitation should not only be restricted to accommodating the homeless but also in restoring dignity to their lives by restoring their means to livelihood.

Agriculture is an important business in Nepal and the farmers would be the ones who have suffered the most in this tragedy. They have lost their homes, family and their farm lands. One of the objectives of the government should be to resuscitate the farming in the area. Although a hard task, the country can borrow ideas from India, as we went through a similar tragedy in 2001 when an earthquake of massive proportions shook Gujarat especially Kutch and Saurashtra regions. Timely intervention by experts like Dr. M.H. Mehta, the then Vice Chancellor of Gujarat Agricultural University helped the farmers to reinvent their profession and helped them to live their lives with dignity. 'Kutch Model' can be emulated in Nepal considering the familiarity that already exists with respect to Green House Technology.

The temporary and permanent shelters that become the shelter for the displaced during the earthquake can be later transformed into very good sources for livelihood for people to grow vegetables, fodder, agroforestry planting material etc. In most situations like these, where people affected are from the rural areas, agricultural should be recognized as the priority area for rehabilitation. These kinds of agricultural activities can be planned and implemented which not only aids in empowerment but also help in attaining self-sufficiency. The agricultural rehabilitation has to be worked out by considering the local situation - crop patterns, availability of seeds and other agro inputs. Massive immunization drive should be carried out for the Livestock. A comprehensive program including Biocontrol agents, low cost green houses, mixed cropping system can be drawn out and given as a package to groups of farmers. A complete package of Eco-Agri / Organic Farming can be adopted for such units. Biocontrol, Bio Composts and Bio Fertilizers packages can be adopted quickly and the extra produce can be marketed under a common brand (organic) for marketing. An integrated farming model – involving Agri – Horti – Livestock emerges as the most optimum model for small / marginal farmers. It may also be

(*Contd.*)

viewed that a catastrophe like this can also give an opportunity to come out of inertia, generate new ideas and fresh thinking and planning and introduction of new technology on a large scale.

As the post Kutch experience and similar other projects such as - Bioshield (Post Tsunami in Kalpakam – Tamil Nadu), and Uttarkashi (Post Landslides) – have proved, several new and useful developments can emerge in these situations which can stay for the greater good of the community in the future. Low cost Drip Irrigation System can be popularized in these areas as they begin rebuilding their water supply systems and irrigations canals. The community can follow Sustainable Integrated Agriculture model and Eco Agriculture i.e., low cost Agri Bio Inputs – The 20 : 20 model of higher Farm Production that can maximize production without compromising on the ecological balance. Agro Waste / Farm Waste can be recycled back in to the system as Bio Composts through Multi-MicrobialTechnology. Similarly, Seed Ball Technology i.e., pelleting the seeds with Agri Bio Products to different patterns of hilly areas to prevent land slide, greening, etc. can be another innovation the hilly terrains of Nepal can experiment with. Co-operative Agro Processing Models can aid the farmers in postharvest operations in securing better economic dividends. (These details are discussed in Dr. Mehta's article "The Kutch Model" – for Rehabilitation of Earthquake affected people in Nepal – June 2015 issue of Agriculture Today as well as the Booklet The Agriculture Rehabilitation in Earthquake affected Gujarat by Dr. M.H. Mehta and Prof. M.S. Swaminathan, GAU Publication 2002).

India has displayed stupendous solidarity with Nepal during the harrowing times. As an empathetic neighbour we can still help them in their rehabilitation process by lending our experience and experts and take the camaraderie to the next level.

Ref : Editorial, Agriculture Today, July 2015

Though for different regions rehabilitation programs may be different the experience gained in Kutch can be a good model to apply in areas like Nepal and other places (now Indonesia also) Depending on the local situation, the design and details were specifically worked out and given to NGOs, Religious Trusts and Government Bodies (Fig. 2).

What is more, it is possible to convert a major catastrophe into an opportunity and to take a quantum jump in introducing improved technology, planning eco agriculture and the Livelihood program for the well being of people. The experience and technology model is available to any organization who would like to help our sisters and brothers in need. We would be very eager to share and support.

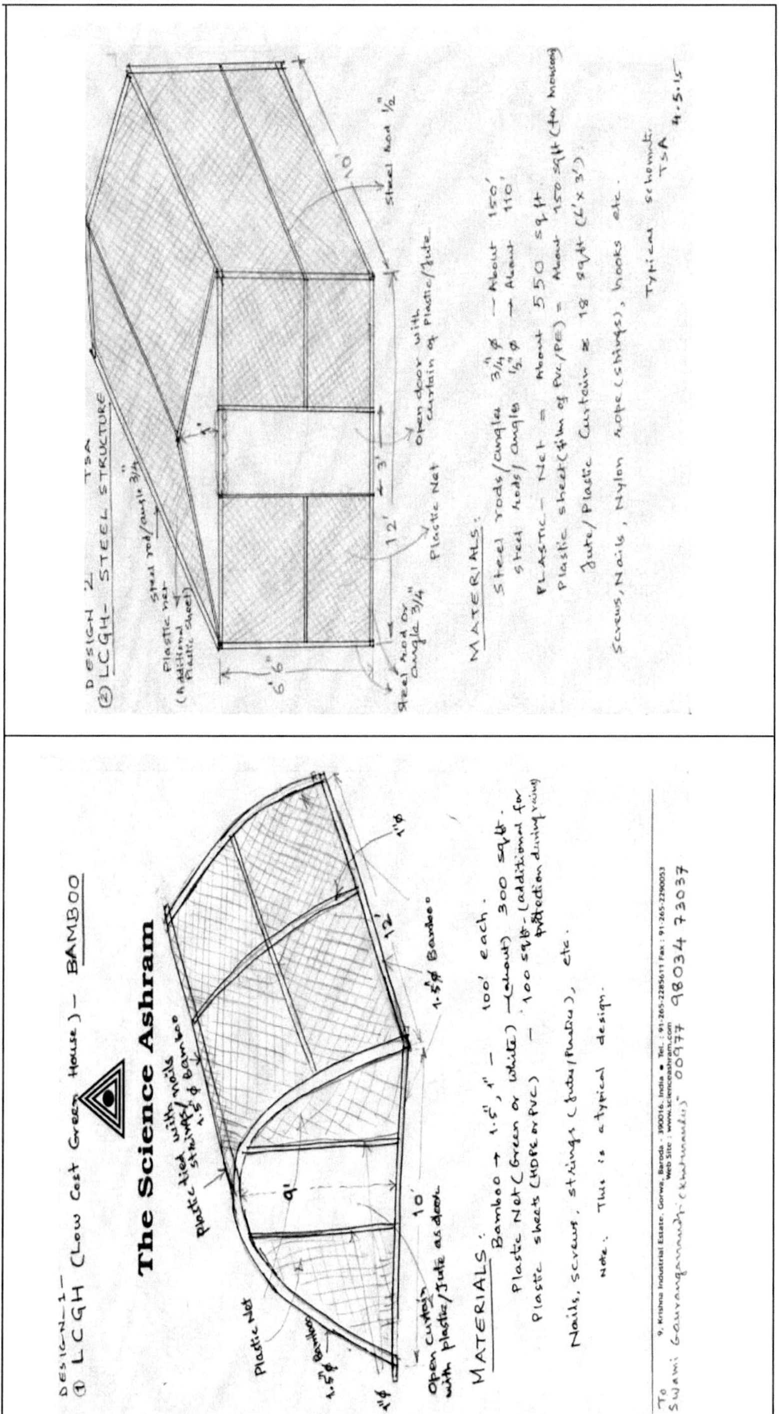

Fig. 2: Two Model Designs

ACKNOWLEDGEMENT

The guidance and support of Prof. M.S. Swaminathan and Dorabji Tata Trust was a great strength. The dedicated work of a large number of volunteers, students and local farmers of Kutch was exemplary in converting a big calamity into an opportunity.

In calamities the creativity, team work and compassion often rise to new heights. As Vivekananda has said, "A few heart-whole, sincere and energetic men and women can do more in a year than a mob in a century".

REFERENCES

1. Mehta M.H. &Swaminathan M.S. – Agricultural Rehabilitation in Earthquake Affected Gujarat -GAU Book publication - 2002
2. Editorial comments – Agriculture Today – July 2015.
3. SwaminathanM.S, "From Rio de Janeiro to Johannesburg – Action Today and not just promises for tomorrow' East West Books (Madras), 2002.
4. Personal Discussions – Embassies of Nepal and Indonesia – 2015 and 2016.
5. Abhijit Banerjee and Esther Duflo, "Poor Economics" Random House (I), 2014.

The creation of a thousand forests is in one acorn.
— Ralph Waldo Emerson

16

Bioshields - The Urgent Need

Bioshields offer barriers of vegetation, shrubs, small and big trees and if integrated as a well designed system they can greatly help our coastal as well as inland areas. Though primarily looked at for coastal regions, under a broader definition Bioshields can play important role in semi-desert areas, hills, etc. also.

WHY BIOSHIELD?

Bioshields or Biobarriers can make a major contribution in minimizing the effects of storms, cyclones and Tsunamis and at the same time offer short and long term ecological and livelihood benefits.

- Bio-Shields in coastal areas include raising mangroves, casurina, palms, bamboo & other tree species and halophytes like Salicornia, etc. followed by other regular plants for the land.
- They serve as speed-breakers under conditions of coastal storms, cyclones and tsunami.
- In addition, they serve as carbon sinks, since they help to enhance carbon sequestration and sustainable fisheries and agriculture.
- Integration of Bio-shield with livelihood options and eco-restoration by developing different site specific models is therefore very important.

For example- the case of two villages near Kalpakkam in Tamil Nadu is often taken. During the devastating Tsunami sometime back, the village with plantation- a natural bioshield was saved with hardly any loss of life as against a nearby village with no natural bioshield. Several such examples are available world over. This indicates that if we can have a good, scientifically designed bioshield, it can be of a great help. This concept is bound to get much greater emphasis when we consider the increasing effect and danger due to climate change in the coming years.

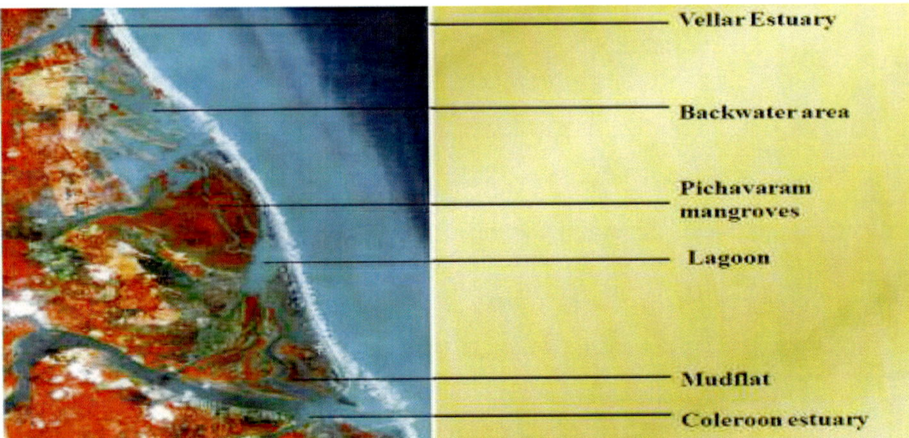

Photo – 1: Remote Sensing Imagery – For Kalpakam

DIFFERENT SALINITY ZONES IN AN ESTUARY

Zone	Range of Salinity
Eulhaline zone	30 and 40% and more
Polyhaline zone	18 to 30%o
Mesohaline zone	5 to 18%o
Oligahaline zone	0.5 to 5%o
Limnatic condition	0.5%o (Freshwater)

Photo – 2: Fish bone type canal as dug in one of the restoration sites in Pichavaram mangrove wetland of Tamil Nadu

1. COASTAL BIO SHIELDS DESIGN

SYSTEM DESIGN – A SCIENTIFIC APPROACH

For many years the role of mangrooves, casurina, etc. has been discussed and several praise-worthy attempts have been made. It is however, very important to have a systematic and scientific design of Bioshield. Broadly this has to be on the following guidelines :

1. **Location specific:** You cannot grow mangroves or casurina in all locations. There have been reports about monoculture of Casurina adversely affecting sand dune ecosystems.
2. **Biodiversity:** Biodiversity is the key for designing a system where suitable shrubs, grasses, small and big trees have to be included for overall benefits.
3. **Role of Halophytes:** Salinity ingration and saline and sordic soils are the major problems of coastal area. On the other hand there are hopeful signs of work on several salt tolerant and economically attractive plants-halophytes like Salicornia, Sueda, Salvadora, etc. which can offer a "win-win" situation both for natural and human habitations. Examples of large plantations of halophytes (mainly Salicornia) in Eritrea and Mexico by Sea Water Foundation, Arizona are inspiring. Through GSFC Science Foundation, The Science Ashram and Navsari Agricultural University several demonstration plots in South Gujarat and Kutch (North Gujarat) have given good results.
4. **Design of Sand Dunes and Boulders:** Sand dunes play a very important role in the coastal systems. A good engineering design of sand dunes with proper angle of repose, coverage with grasses and vegetations is important. In several critical areas, boulders have to be created such as stone walls, concrete fence, etc.

DEMONSTRATION PROJECT AT KALPAKKAM

During the Tsunami, as in many other places the Indira Gandhi Centre for Atomic Research (IGCAR), Kalpakkam in Tamil Nadu, a great deal of damage and loss of human lives occurred. Soon after the Tsunami, the concept of a scientifically designed Bioshield was presented by The Science Ashram- an NGO in Baroda-India.

In fact, during one of my earlier visits to IGCAR Kalpakkam, forgiving a memorial lecture, I had discussed the need for a Bio Shield in that area and later it.

It was decided to set up a Bioshield at Kalpakkam through a consortia of groups / institutions like The Science Ashram, MSSRF (MS Swaminathan Research Foundation- Chennai), Navsari Agricultural University, Gujarat Life Sciences, CSMCRI-Bhavnagar and IGCAR.

Based on the local conditions surveyed, designs were prepared for 3 locations for demonstrations of Bioshield. The major considerations are as per the details in Table-1.

Table 1: Survival and growth in various salt tolerant fruit bearing species at Site- 2

Sr.No.	Name of fruit bearing Spp.	No. of Plants	Height (Cm.)	No. of branches	Remarks
1	*Salvadoraoleoids*	132	182	10	-
2	*Colophylum inophyllum*	40	138	4	-
3	*Dilleniapentagyna*	20	75	-	-
4	*Syzygium* spp.	17	138	6	In fruiting
5	Sapota (*Achras zapota*)	18	135	-	In fruiting
6	*Punica granatum*	17	82	5	In fruiting
7	*Emblica officinalis*	8	204	-	-
8	*Annona squamosa*	7	105	-	In fruiting
9	*Psidium guajava*	4	130	5	In fruiting
10	*Simarouba glauca*	5	105	-	-
11	*Zisyphusmauritiana*	1	200	-	In fruiting
12	Citrus (lime)	5	115	-	-

BIO SHIELD DEMONSTRATION SITE - KALPAKAM

In Photograph 1, 2 & 3 the scientific representation of a Bioshield Demonstration is given. The Bioshield is conveniently divided in to 3 Zones :

Salicornia (1)

Bio Shield Demonstration site - Kalpakam (2)

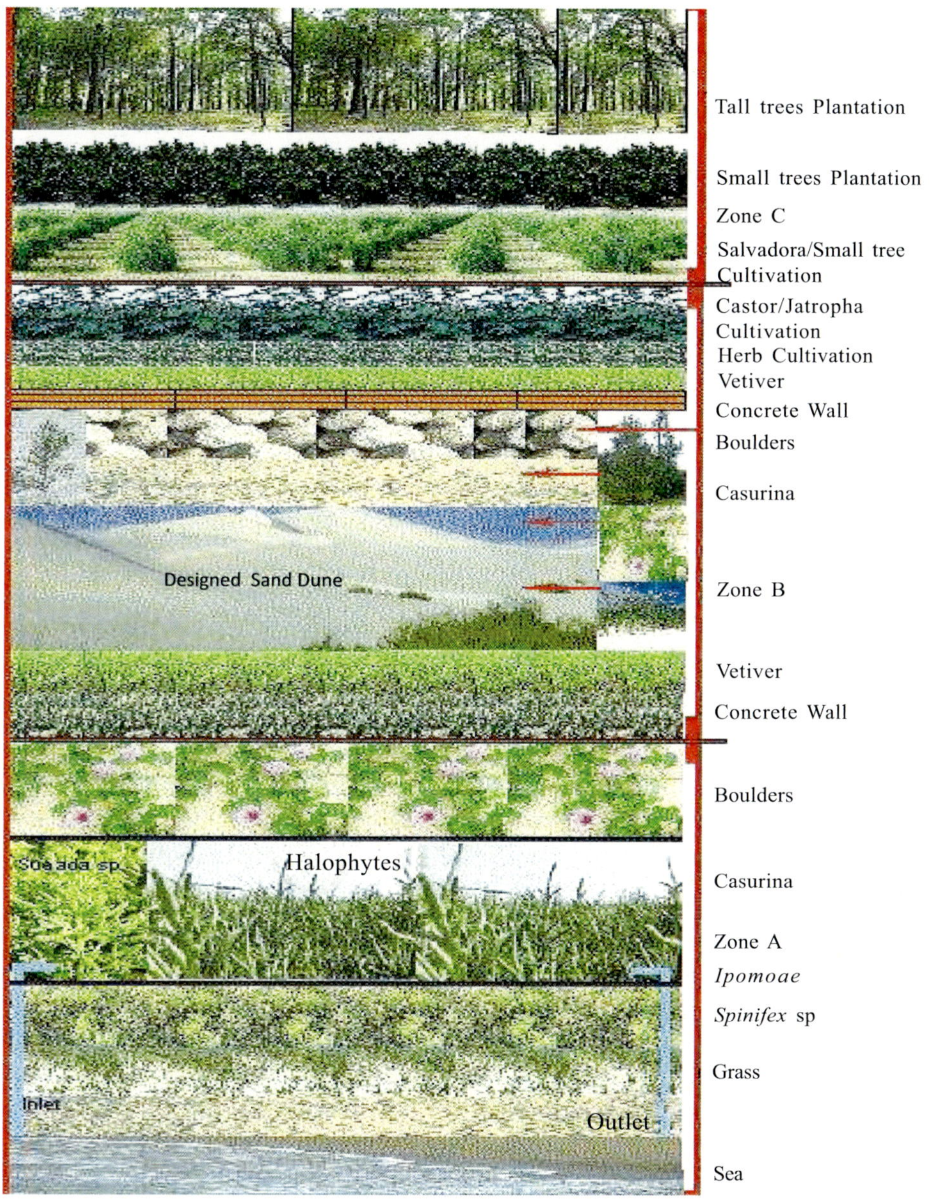

Photo – 3: Biosheld Demo Project (Kalpakkam)- Schematic Design

Zone-A Grasses, Spinifex, ipomoea, Salicornia, Sueda and other halophytes (Mangrooves in some locations).

Zone-B Designed sand dune, Boulders, Concrete Wall, cultivation of non-edible oil plants, vetiver grasses, etc.

Zone-C Medium to tall trees like Salvadora, Casurina, etc.

BOX - 1
BIO SHIELD COMPONENTS

Halophytes

Sea water agriculture is the strategy to link ecology and livelihood. Halophytes are salt tolerant plants. Halophytes such as *Salicornia brachiata, Sueadamaritia* will be irrigated with salt water on a pilot scale and standardization of technology will be made for replication.

Sand Dunes

Sand dunes are designed as natural buffer to the coastal system; they protect the regions behind them from wind and waves, and provide a substrate, or ground, for plants to form communities upon. Plants like grasses, *Ipomoea pes caprae*, Vetiver, *Spinifex* sp etc., will be grown as vegetation cover

In natural conditions wind continues to move sand up to the top of the pile until the pile is so steep that it collapses under its own weight. The collapsing sand comes to rest when it reaches just the right steepness to keep the dune stable. This angle, usually about 30-34°, is called the *angle of repose*. Therefore, the artificially raised sand dunes will be created with an angle of repose around 30-34°.

Grasses

Local grasses such as *Cynodon, Paspullum* and including other salt tolerant palatable species found in Kalpakkam area will be identified for dune stabilization.

Non Edible Oil Plants

Non edible oil plants such as Castor, Jatropa, Vetiver, Neem, *Salvadora* and *Calophyllum inophyllum* will be planted in the bio shield zones for commercial returns and ecological improvement.

Medicinal Plants

Plants with medical importance and of commercial returns like *Vincarosea, Aloe vera, Clerodendrum inerme* (reported as substitute for quinine), *Vitex trifolia, Morinda citrifolia* etc., will be planted

Timber and Trees with Aesthetic value

Trees which offer timber, *Simarouba glauca*, Casurina, Coconut, Palm, Bamboo, *Thespesia*, Banyan, *Sapindus*, *Acacia auriculiformis*, Neem, *Anacardium, Spathodea* etc will be planted. These trees would offer both ecological and economic benefits.

The established seedlings of indigenous trees, *viz. Tamarindus* indica, *Pongamiapinnata* and *Erythrina* indica have their ability to grow on the sand dunes.

Rock Plants

Drought and salt tolerant xerophytes such as *Cactus, Acacia, Euphorbia, Tillandsia, Nerium oleander* etc will be include in the artificial boulders establishment. *Ficus* trees will be also introduced for its evergreen, high anchorage and drought tolerant ability.

THE PROGRAM

Program for Bio Shield Project

The stepwise program involved

- Establishment of nursery for raising of seedlings required for plantation at sites.
- Procurement of samplings of salinity tolerant fruit and other trees.
- Procurement of seeds of Salicornia, Castor, Simarouba, Jatropha, Senna, etc. for direct and nursery sowing.
- Preparation of nutrient rich biocompost .
- Plantation of Salvadora, Simarouba, Castor and Jatropha at Site- 1
- Plantation of fruit bearing trees and halophytes *viz.,* Salicornia etc. at Site-2.
- Plantation at Site- 3.
- Awareness programme

Establishment of Nursery

To meet the requirement of planting material a nursery was established in the IGCAR Township. Polythene bags filled with sand, soil and FYM in equal proportion were used in beginning and later on Superlife and Tricholife were added to nursery mixture and FYM was replaced with biocompost. This mixture enhanced the germination, growth and survival of the seedlings. The establishment of nursery was very successful and we could produce sufficient quantity of saplings of *Salvadora oleoides*, *Jatropha carcus, Simarouba glauca* etc.

Procurement of saplings

Besides raising of seedlings in nursery, to meet the immediate requirement for Site -2 saplings of Sapota (*Achraszapota*), Custard apple (*Annona squamosa*), Guava *(Psidium guajava*), Pomegranate *(Punica granatum),* Bel *(Aegle marmelos*), Jamun (*Syzygium cumini*), Aonla (*Emblica officinalis*), etc., were procured from Chennai based nursery . Mr. P K Biswal Project Officer managed saplings of Tendu (*Diospyros melanoxylon*), Malayapple (*Syzygium* spp.) etc., from Odisha. To raise Salicornia, saplings were brought from Pichavaram and nearby seacoast of Mahabalipuram.

Procurement of quality seeds

TSA arranged quality seeds of Pilu (*Salvadora oleoids*), Ratanjayot (*Jatropha carcus*), Simarouba (*Simaroubaglauca*), Custard apple (*Annona squamosa,*) Senna (*Cassia angustifolia*), Kumet (*Acacia mangium*), Tumba (*Citrullus Colocynthis*), Castor (*Ricinus communis*), Salinity tolerant paddy, Salicornia, etc. from NAU, SDAU and CSMCRI for raising of seedlings and direct sowing.

Preparation of nutrient rich biocompost

Sufficient quantity of undecomposed farm yard manure (FYM) was arranged from nearby villages for production of biocompost. FYM was initially sprayed with wonderlife-1 and thoroughly mixed . After 15 days FYM was sprayed with wonderlife 2 &3 and thoroughly mixed. The mixing was repeatedly done at an interval of 7days and sprayed with water for about two months. The microbial application enhanced the decomposition rate and made the nutrients in available form. The biocompost so produced was used in filling of polythene bags and pits for plantation of fruit and other tree saplings.

Typically the width of such a Bioshield may be 300 to 400 mtrs. Such a project can serve as demonstration facility which can be repeated in many more locations with modifications / changes as per local conditions. Some relevant photographs show the progress and results of field experimentation.

Coastal Power Plants – Desalination Plants and Bioshield

The present trend of setting up Nuclear and Thermal Power Plants in the coastal area and combining them with water desalination plant is likely to increase further. This is because power plant - desalination combination systems are comparatively more economical and reliable. Desalination will become increasingly popular with improved membranes. Combining Bioshield system with the desalination plant can be an attractive system to take care of the concentrated saline water stream and growing of halophytes like Salicornia. Such a typical model is shown in Photograph – 4.

Bioshield system can even be combined with inland brackish water desalination plant such as in the semi desert areas of Rajasthan, Andhra Pradesh and other places. The concentrated brackish water in such cases becomes the feed for the halophytes. Such units are being planned in Barmer District of Rajasthan. Similarly Aqua-agro system in the coastal area can be set up in combination with Bioshield system with much advantages. The concept of Bioshield can be

Photo – 4: Design Models for Integrated Bioshield

Typical Model -1: Aqua Agro Bio-Belt Design

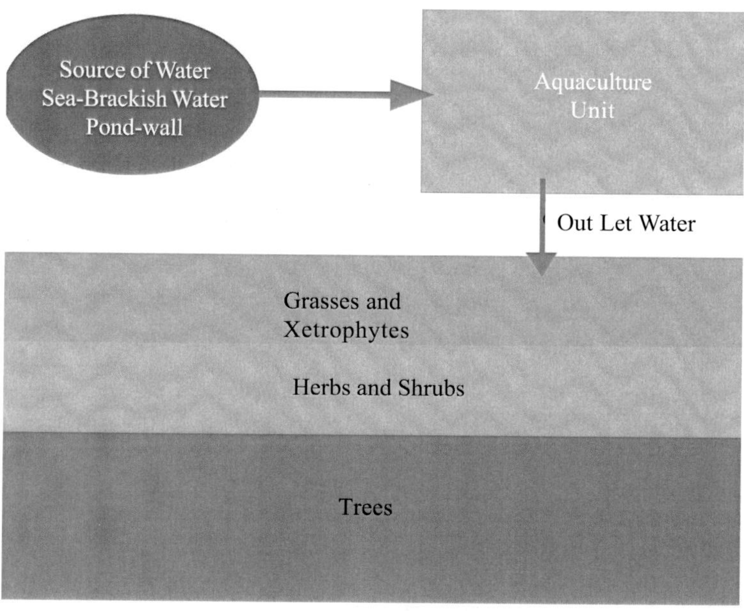

The Science Ashram Vadodara

Typical Model – 2: Bio-Belt Design for Coastal Area

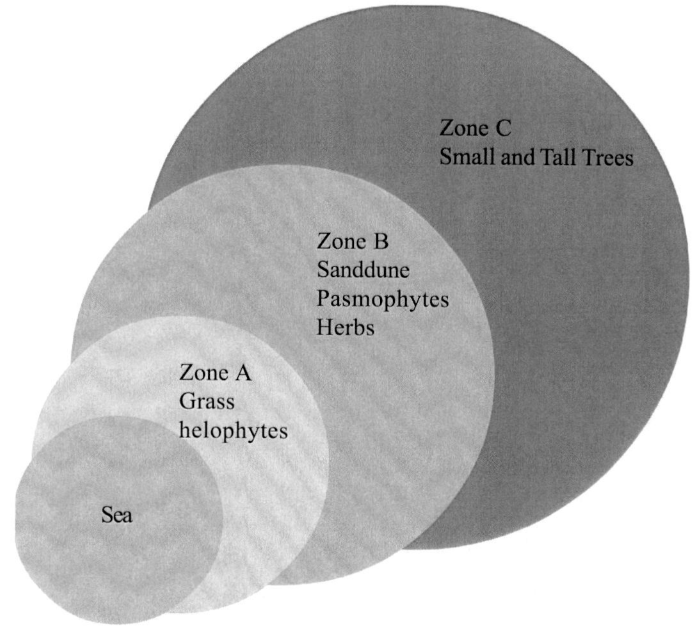

The Science Ashram, Vadodara

Typical Model -3: Bio-Belt Combined with Desalination Plant

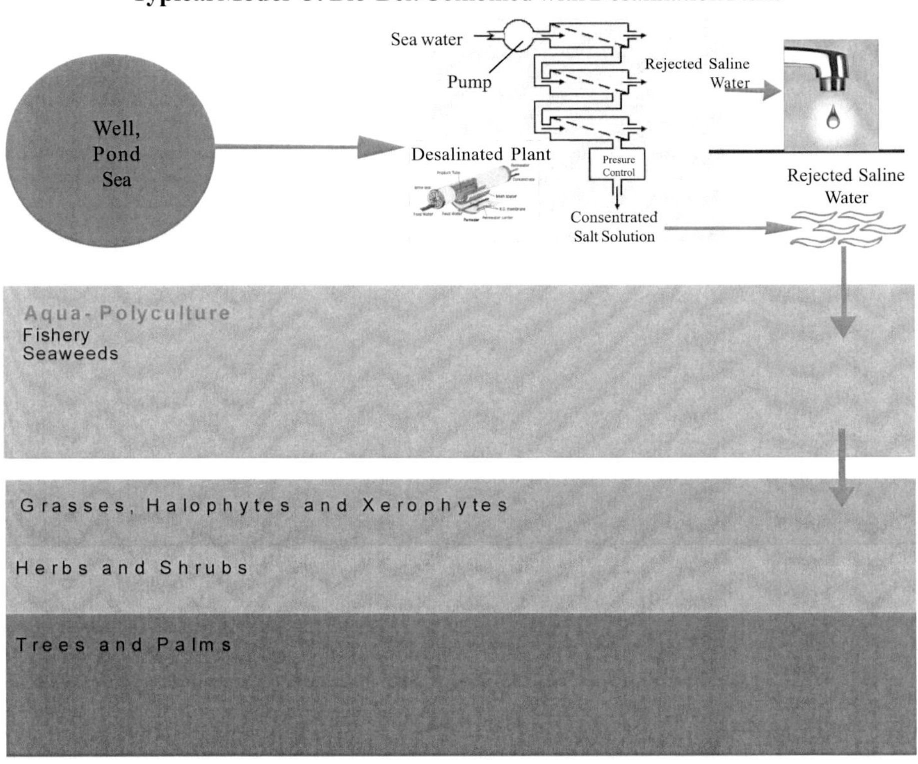

The Science Ashram, Vadodara

broad and applicable for hilly areas as well.

Taking a leaf from such a demonstration of bio shield at Kalpakkam, a recommendation was prepared for Atomic Energy Commission to include Bio shield as an integral part of Coastal Power Plants.

Similarly, for Government of Gujarat a proposal was prepared (BOX - 2). It is now being included as a part of broader proposal for prevention of coastal erosion as the 'soft' alternative.

BOX – 2

Economic Times – 21st November-2008

Bio-shield protection sought for state's 1,600 km coastline

Yashpal Parmar
VADODARA

AFTER the successful implementation of the bio-shield project at the nation's top security ranked assets—Indira Gandhi Centre for Atomic Research (IGCAR) at Kalpakkam—now the city-based Science Ashram has proposed to the Gujarat government to consider bio-shield protection for its 1,600 km coastline, the longest in India. Bio-shield, divided into three zones, protects coastal settlements, including industries, from natural disasters, offers economic returns and earns carbon credits as well.

Science Ashram, an NGO engaged in educating farmers and promoting agri biotechnology, has designed a special scientific three-tier bio-shield for IGCAR. This three-year project is being implemented in tie-up with MS Swaminathan Research Foundation, Navsari Agricultural University, Gujarat Life Science and IGCAR team. IGCAR is now a national demonstration site owing to its scientific and systematic implementation. Science Ashram president M H Mehta, also former vice-chancellor of the erstwhile Gujarat Agricultural University, told ET, "I have written to chief minister Narendra Modi and we have received positive response as minister for environment is considering a personal visit for this project demonstration." It is designed in such a way that apart maintaining bio-diversity, the bio-shield provides vital protection from cyclone, Tsunami, and other disasters and in turn, offers more green cover and prevents tidal waves washing away coastal areas reducing soil erosion, Dr Mehta said.

"Currently, we are busy in the assembly elections, but we will definitely consider it as and when it gets finalised," environment minister Mangubhai Patel told ET. "Bio-shield, zone A consists of varieties of halophytes like salicornia, sueda & mangroves depending on respective locations that withstand salty waters, zone B has designed sand dunes, casuarinas & shrubs-herbs plantations including boundary wall as barriers while zone C will have medium & large trees offering green cover and absorb the greenhouse gases," Dr Mehta expounded. Industrial houses with operations in coastal areas can take a lead role in popularising and promoting this bio-shield for the Gujarat coastline, Dr Mehta told ET, adding that carbon credit, which is gaining importance nowadays can prove major boost to push in this direction as well.

Gujarat can start with small demonstration sites and then can expand into different areas. Even Vadodara's River Vishwamitri, given its unique bio-diversity, can be equipped with bio-shield and make it pollution free and more environment friendly, he said.

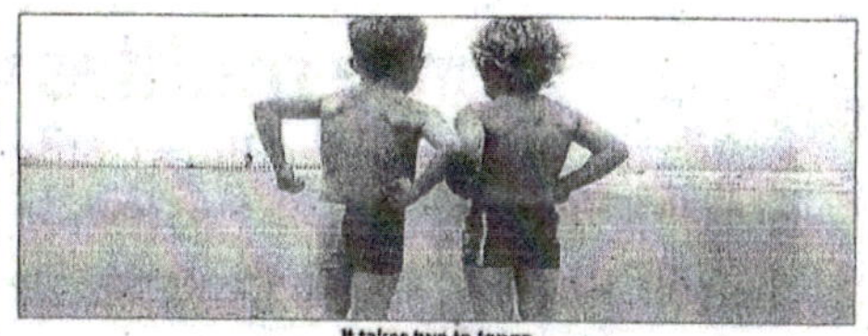

It takes two to tango

RIVER REVIVAL & BIOSHIELD

Bioshields should also be an integral and important part of river revival system. After all, each tree and shrub on the bank of a river acts as a 'micro check dam' as discussed later in a special chapter 'Vaho Vishwamitri Abhiyan – River Revival'. The experience of Kalpakkam greatly helped in planning of Bioshield for this river revival movement. The scientific planning of Bioshield, varities of shrubs and trees should become a major component of a holistic plan for the river revival.

BIOSHIELD FOR PROTECTION OF HILLY AREAS – HIMALAYAS

In the earlier chapter on farm residue waste and bio compost making the case study for pine needle waste was cited. One major reason cited for landslides and flash floods in areas like Uttarkashi – in Himalayas was the loss of vegetation and bio diversity of the region.

Based on several visits and discussions with the local NGOs like SEED (Science for Equity, Empowerment and Development), it was decided to go for a Bioshield Demonstration project at 2 or 3 locations near Uttarkashi.

BOX – 3

TOI, 4.9.13

Guj shield to protect Uttarakhand

Project Designed By City-Based M H Mehta Will Help State Tackle Natural Calamities

Prashant Rupera | TNN

Vadodara: This project prepared in Gujarat will provide a shield to flood-ravaged Uttarakhand. The National Bioshield Society (NBS) headed by city-based Padma Shri M H Mehta has prepared a scientifically designed project to help Uttarakhand get a natural shield against floods, especially its vulnerable plains.

The first phase of the project, which will focus on Uttarkashi and Gangotri regions, has been cleared by the Central government after the project proposal was discussed during a meeting chaired by R Chidambaram, principal scientific adviser to the Union government and chairman of the scientific advisory committee to the Union cabinet.

It is worth mentioning here that it was due to Mehta's efforts that Kalpakkam in Tamil Nadu got bioshield protection after tsunami had hit the Indira Gandhi Centre for Atomic Research on country's coastal belt.

"For developing a bioshield system in Uttarakhand, a consortium is being planned. This consortium will include members from Amrita University, Ministry of Earth Science among others," Mehta, former vice-chancellor of Gujarat Agriculture University (GAU), told TOI.

Experts believe that the devastation due to flash floods at Uttarakhand was inevitable with the artificial forestation that the region witnessed from the time of British rule.

This artificial forestation killed the ethnic biodiversity of the region and instead developed a monoculture which focused on pine forests that proved very profitable for both timber and resin.

"Pine created a smooth, dry carpet, which absorbs nothing, and which at the same time prevents the development of any undergrowth of other plants. In fact, the ground in a pine forest is as bare as a desert and in Uttarakhand it has even turned the soil acidic," said Mehta.

Mehta on the advice of world renowned activist and Padma Bhushan Sundarlal Bahuguna, the octogenarian who founded the Chipko movement, had started working on the Uttarakhand project proposal.

"Now, through the consortium, we are preparing a demonstration model which can minimize the landslide problems. There should be a bioshield of herbs, shrubs and trees suitable to the region and that can regenerate biodiversity. This should act as a model for many cases in other parts of the country and the Himalayan region," he said.

Also, NBS is working on another project to convert the agro waste at Uttarakhand into bio-compost.

"The pine waste is not only spoiling the soil and turning it acidic but also results in forest fire. To collect the waste and convert it into good quality bio compost which is anyways in big demand, we are working with Seeds India, an NGO comprising of young members in Uttarkashi," said Mehta, chairman of the regional Community Science Centre.

In order to speed up the plantation and to reach difficult areas, 'Seed Ball Technology' developed was taken up. Details of Seed Ball Technology is discussed elsewhere. As soon shown in Photograph, seed ball camps for plantation were taken up by children and local women with much enthusiasm. Selection of a variety of appropriate seeds and coating them with a mixture as shown in Fig. 1.

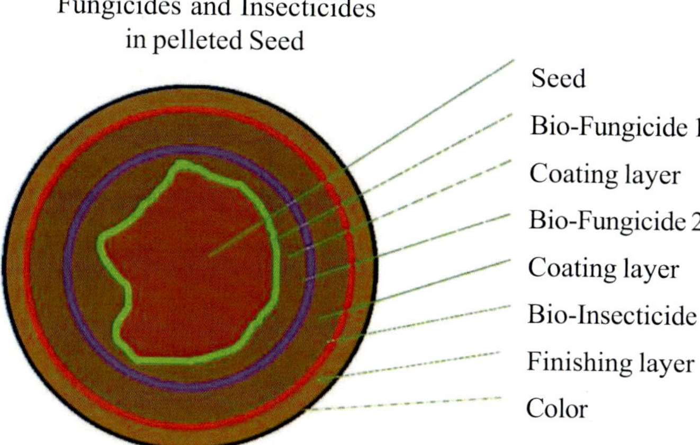

Fig. 1: Pelleting and coating seeds – Seed Ball

Seed Ball Camp

Advantages like ability to reach difficult terrains, preservation of seeds and higher success rates can be achieved through this. (More details about Seed Ball Technology is discussed in a separate chapter – Vaho Vishwamitri Abhiyan – River Revival).

Such camps were also organized in response to other organizations like 'Green Goverdhan' and 'Yamuna Bachao' movement near Mathura - UP

BIOSHIELDS-NEED FOR A NATIONAL PROGRAMME

There is a strong need to take up an integrated National Programme for Bioshield. The promotion of coastal forestry and agro-forestry will confer both short and long term ecological and livelihood benefits and at the same time offer a shield against natural disasters.

Carbon sequestration through coastal Bioshields will make an important contribution to promoting a balance between carbon emission and absorption, in addition to offering protection during coastal storms and calamities like Tsunami, cyclones, etc.

It has been frequently discussed by many and emphasized by Scientists like Dr. MS Swaminathan that "No further time should be lost in initiating a national coastal bioshield movement along the coasts of the mainland of India as well as islands." It is also going to be necessary to plan such a programme through scientific system/design approach and keeping overall factors in mind as being demonstrated at Kalpakkam and other places. Based on sound and scientific approach and plan, such an implementation can be taken up through people participation. Bioshields will play important and integral role in holistic and eco friendly development.

REFERENCES

1. The Science Ashram Report on Bio Shield – Coastal Demonstration Project at IGCAR, Kalpakkam, Tamil Nadu, October 16, 2010.
2. Bhalla R.S., "Do Bio Shields affect Tsunami inundation? Current Science, Vol. 93, No. 6, 25th September 2007.
3. Mehta M.H., Landslide Prevention Measures, Brain Storming, Office of Principal Scientific Advisor, GOI, Vigyan Bhavan, New Delhi, 12th April 2012.
4. MSSRF, Chennai Annual Report 2009.
5. Carl N.Hodges et al, Reversing Flow : Water & Nutrients from Sea to Land; AMBIO; Vol. 22, No. 7, Nov. 1993.

"Jin Khoja Tin Paiya Gahre Pani Beth" ("one who searched in deep waters found it")

— *Kabir*

17

Saline Water Agriculture

Saline Water Agriculture has been drawing interest as one of the solutions for limitation of world fresh water supply, farmland availability, salinity ingression and the pressing future need of food supply. The cultivation of salt-tolerant crops on saline soil and with saline waters has significant social and economic potential.

It is often argued that fresh water agriculture near rivers and ponds was invented because of the need to sustain people and communities. Saline water agriculture is going to be needed for sustaining our future generations and civilization.

FROM DESALINATION OF SEA WATER TO SEA WATER AGRICULTURE

Sea water contains nearly 35000 ppm of dissolved salts. There has been great deal of interest world wide on desalination of sea and brackish water to safe drinking water. In early 1960s the young President of USA John Kennedy put forward two famous challenges for the scientists (1) Landing the man on moon and (2) Producing desalinated water at low and competitive cost. It is often said that the first challenge was met but the second one has been more difficult!

With tremendous support and interest, work on desalination was taken up worldwide. Newer processes were being tried. At the Environmental Research Lab of University of Arizona and at Central Salt and Marine Chemicals Research Institute, Bhavnagar simultaneously solar powered HD (humidification – dehumidification) process of Desalination was developed and pilot plants were commissioned. It was, however, realized that desalinated water was costly and not easily applicable to most coastal communities, though for many countries in middle east with no alternative water source and cheap petroleum energy – it is applicable on a very large scale.

A radically different approach was developing technology for using saline water directly for agriculture.

In the final year Chemical Engineering Student at I.I.T Bombay in 1966, I was given a project – Design of Desalination Plant for the brackish waters of Rajasthan. In this, the design of a solar powered H – D process was prepared. A great deal of back up information provided by the Environmental Research Lab (ERL), Arizona and CSMCRI – Bhavnagar reports were helped in design of the plant. Later, after graduation, I joined the CSMCRI team at Bhavnagar to set up a 1000 gpd pilot plant of Solar powered H – D Process. In the same institute another group has started research project on 'Sea Water Agriculture'. In November 1967, when our H – D Desalination Pilot Plant was commissioned, my friend Carl Hodges – the young and visionary Director of ERL – travelled to Bhavnagar. This was followed by several exciting and innovative exchanges about saline water agriculture. Carl's work at Tucson – Arizona and Sonora village – Mexico on Desalination complex with green house based 'Food factory and saline water agriculture as a part of Aqua – Agro Complex. There was much excitement and dreams when I was invited by Carl to visit and spend some time at the facilities in Tucson Arizona as well as Sonora Mexico in 1972. Carl and his team moved very fast and established Sea Water Farms companies in Mexico, Abu Dhabi and Eritria. The important base for such farms were halophytes!.

HALOPHYTES

Halophytes are among the few species of higher plants that can withstand high salinity without detrimental effects. We can make a broad classification of halophytes on the basis of their salt tolerance capacities.

i) Plants with a high salt tolerance able to grow in sea water

ii) Plants with average tolerance i.e brackish water crops and

iii) Crops with moderate tolerance slightly brackish waters.

About 2600 halophytic species are known and only few are extensively studied for their potential in agriculture and as source of oils, flavours, guns, resins, pharmaceuticals etc. Also their environmental potential for protection and conservation of coastal ecosystems, soil improvement and source for biomass for biofuel etc.

For halophytes to succeed as useful crops, the basic conditions needed are : (a) Good yield (b) proper irrigation (c) growth in existing infrastructure. Some major halophytes of importance are listed here:

1. **As Food and Oil seed:** *Aster tripolium* (sea spinach), *Salicornia bigelovii, Salicornia brachiata, Kochia scoparia, Salvadora oleoids etc. Salicornia* in particular, has been grown in large scale and the seed give

31% protein, 28% oil rich in polysaturated fatty acids (74% linoleic acid) and the tips used as vegetables.

2. **As fuel wood and Timber :** More than a billion people in developing countries rely on wood for cooking and heating. Spices of Prosopis, Salsola, Acacia, Sueda, Pongania Pinata, Avicenula are good fuel woods.
3. **Environmental properties :** Halophytes like *Avicennia marina*, *Spartina maritima*, *Avicennia germinans* (black mangrovs) etc are known to restore coastal area. (we have explained this in detail in the chapter on Bioshield).

The use of halophytes for CO_2 absorption from air is another attractive use.

MECHANISM OF SALINITY TOLERANCE

The reduction in germination and low productivity may be attributed to increased osmotic pressure of the soil solution, which diminishes the water absorption rate leading to moisture stress in seeds. Salt stress interrupts metabolic activity by ion excess, imbalance and interference of toxic ions on the uptake of essential nutrients which are detrimental to seed germination under saline conditions. A better understanding of the underlying mechanism involved in seed germination and the growth under saline conditions is needed to confront this agronomic problem. The cell membranes form barriers between the plant and its environment, and thus membranes are of special importance to plants for regulating the ion content.

PRODUCTION OF HALOPHYTES

Hugo Bayko in Israel drew attention to the use of sea water as an irrigant. Successful Sea Water irrigation depends on the use of Halophytes.

At Central Salt & Marine Chemicals Research Institute (CSMCRI) Bhavnagar, response of sea water irrigation was exclusively studied. Many of the crops varieties tested tolerated upto 10000 ppm saline water without a significant reduction in the germination potential as shown in Table – 1 (Reddy &Iyengar).

Table 1: Nutritive Value of Bajra (Millet) and Wheat Grown with Seawater (percentage constituents per 100 g seeds)[a]

	Bajra (Millet)[b]		Wheat[c]	
Constituent	Seawater	Normal Sample	Seawater	Normal Sample
Moisture, %	7.7	12.4	8.6	12.8
Protein, %	9.5	12.6	17.3	11.8
Fat,%	5.0	5.0	1.5	1.5
Minerals,%	2.1	1.3	1.5	1.5
Fiber, %	1.2	1.2	2.1	1.2
Carbohydrates,%	74.5	67.5	69.0	71.1
Calorific value,%	381.0	361.0	359.0	346.0
Calcium, mg	51.0	42.0	51.0	41.0
Phosphorus,mg	384.0	296.0	248.0	306.0
Iron, mg	20.0	5.0	10.3	4.9
Thiamine, mg	4.46	0.33	0.52	0.45
Riboflavin, mg	0.17	0.25	0.18	0.17
Nicotinic acid, mg	1.8	2.3	4.9	5.5

[a]Samples were tested by the National Institute of Nutrition (ICMR), Hyderabad, India.[b]Direct irrigation with seawater (18,000 – 24,000 ppm) [c]Diluted seawater (10,000 – 20,000 ppm)

Management practices are being developed to enhance the productivity and success rates of germination of Halophytes. There are – Irrigation methods, selection of crops and crop rotations, salt balance and drainage requirement, use of amendments and overall agronomic practices. Gupta and Gupta have emphasized drip irrigation system as well as use of liquid bio fertilizers like Azospirillum and seed treatment with calcium salts.

For the large commercial scale demonstration, the ERL University of Arizona team have reported product uses of Halophytes in different locations and is given in Table – 2 for different species of Halophytes.

Table 2: Productivity of halophytes in managed systems based on experiments by Environmental Research Lab.

Location	Species	Yield t/ha	Soil	Salinity (%)
Safford, AZ	*Atriplex lentiformis*	14.7	Loam	0.6
	A.nummularia	12.3	Loam	0.6
San Juaquin, CA	*A.barclayana*	21.6	Clay	1.0
	A. lentiformis	18.8	Clay	1.0
	A. holocarpa	18.4	Clay	1.0
Tucson, AZ	*S. bicelovii*	17.3	Sand	1.0
Puerto Penasco,Mexico	*A. lentiformis*	17.9	Sand	4.0
	B. maritima	17.4	Sand	4.0
	A. canescens	17.2	Sand	4.0
	S. bicelovii	15.4	Sand	4.0
Kino Bay, Mexico	*S. bicelovii*	20.0	Loam	3.8
Kalba, UAE	*S. bicelovii*	18.0	silt	4.0

Sources :

1) Watson et al. (1987)

2) Watson (1990)

3) Glenn & O'Leary (1985)

4) Mota. C. Unpublished results of 1987 field trials at Kino Bay and 1989 field trials in United Arab Emirates.

Our own experimental demonstration work at three different locations in India – namely Mundra – Kutch, Kalpakkam – Tamil Nadu and Ubhrat – South Gujarat coast have been encouraging. These are shown in the photographs here.

(a) Mundra – Kutch

(b) Kalpakkam – Tamil Nadu

(c) Ubhrat – South Gujarat

Saline Water Agriculture Salicornia Plantation
Mundra-Kutch, 1999

Ipomoae pes caprae
Salicornia spp.
Ipomoae pes caprae
Spinifex sp
Grass
Sea

While doing Halophyte plantation at the experimental farm of GAU (now NAU) near Ubhrat – South Gujarat, I had discussed with our field officers about the 'economics' of the project. It was calculated that if we can get about Rs.3/Kg of Salicornia to be sold as 'salty green vegetable' we will break-even'. Later it was reported to me that it became a popular ingredient for salads as well as for local girls observing a special puja (where eating table salt is forbidden during those days !) and the price of green tips fetched was upto Rs.25/Kg! New varieties of vegetables are always in good demand everywhere !

Comparing these data with the team at Arizona and Mexico of Sea Water Farms & Planetary Design Corporation who reported the cost of cultivation of halophytes as $25/ton for Brackish Halophytes and $ 40/ton biomass for Sea water Halophytes and reportedly profitable as personally discussed with Carl Hodges at that time.

With Carl Hodges – Sonora – Mexico, Sea Water Farm

AQUA AGRO COMPLEX

The ERL team and later the company Sea Water Farms worked out a project proposed for Integrated Aqua – Agro Complex with the following components:

a) Aqua farm consisting of Shrimp and other high value fishes

b) Coastal Halophyte plantation

Further, it was worked out that the effluent saline water from the shrimp farm can be an excellent feed for halophyte plantation. Large scale 'Sea Water Aqua Agro Farms' were set up in Mexico and Eritrea. Later, after several round of discussions and visits a consortia consisting of Sea Water Farms Co. Sea Phire – USA, GSFC Science Foundation, M.S. Swaminathan Research Foundation and Gujarat Agricultural University - GAU was formed to work

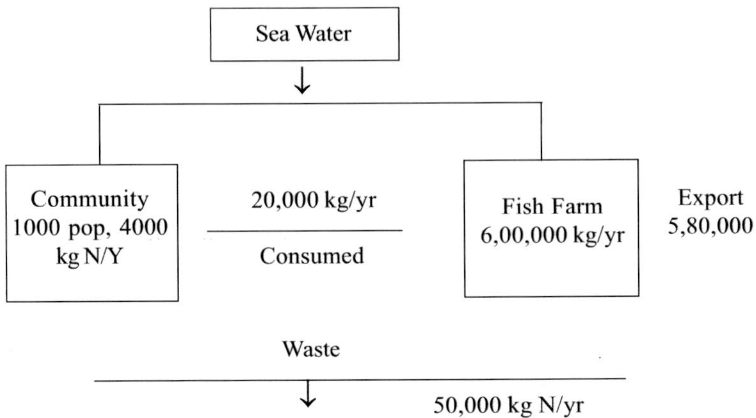

Fig. 1: Sea water Aqua Agrofarms

together for Aqua Agro Complex including Sea Water Agriculture. A typical Aqua Agro Complex proposed is depicted in Fig. – 1

There is good profitability in shrimp production and when the effluent from the pond is used to feed the Halophyte farms and there are several additional advantages as indicated in Fig.- 2.

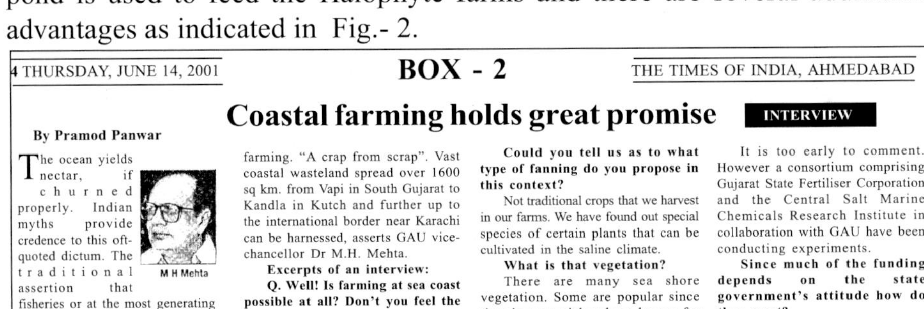

4 THURSDAY, JUNE 14, 2001 **BOX - 2** THE TIMES OF INDIA, AHMEDABAD

Coastal farming holds great promise

INTERVIEW

By Pramod Panwar

The ocean yields nectar, if churned properly. Indian myths provide credence to this oft-quoted dictum. The traditional assertion that fisheries or at the most generating wind based hydropower project can be expected on the sea shore, has not only been belied by the scientists but also explored new vistas of seashore farming higher to alien in conventional farming.

M H Mehta

Recent experiments successfully carried out by Gujarat Agricultural University (GAU) could bring about a revolutionary change in the traditional concept of farming. It is rather incredible to think of any farming under the salty water of the sea. We can harvest much needed valuable foreign exchange worth millions by sea shore farming. "A crap from scrap". Vast coastal wasteland spread over 1600 sq km. from Vapi in South Gujarat to Kandla in Kutch and further up to the international border near Karachi can be harnessed, asserts GAU vice-chancellor Dr M.H. Mehta.

Excerpts of an interview:

Q. Well! Is farming at sea coast possible at all? Don't you feel the very concept is next to impossible?

A. 1, too. thought alike but the recent experiment conducted at the coastal land of Mundra and Mahuwa have amazed every one. Yes. We could cultivate profitably along the coast without desalting the water. There is no need of desalination of the seawater, whatsoever!

Don't you think the cost will be exorbitant?

No! We are not for desalination of the seawater. Desalination device is costly: on the contrary the new method will be cheaper as we would use water which is not used for drinking or irrigation.

Could you tell us as to what type of fanning do you propose in this context?

Not traditional crops that we harvest in our farms. We have found out special species of certain plants that can be cultivated in the saline climate.

What is that vegetation?

There are many sea shore vegetation. Some are popular since time immemorial and are known for herbal and medicinal value. We have however taken up *Salicornia*. used as a popular food component, in western countries.

What is appealing and special about *Saliaimia'!*

What is the most appealing aspect is that it does not need the traditional irrigation water, which oil seeds and fodder crops do. With minimum cholesterol content *Salicornia* has become a safe food habit. More over promise its bio-product has proved as an ideal fodder for the cattle.

How do the experts take it?

It is too early to comment. However a consortium comprising Gujarat State Fertiliser Corporation and the Central Salt Marine Chemicals Research Institute in collaboration with GAU have been conducting experiments.

Since much of the funding depends on the state government's attitude how do they react?

The state government, indeed, has a positive approach. Chief minister Keshubhai Patel has taken interest in the project and had discussions with me.

As a large coastal land under the project comes under earthquake affected region, do you think it will in some way benefit the affected people?

Yes! I feel the project may have special relevance where GAU is working with Dr Swaminathan Foundation for the rehabilitation of the quake-affected villages in Kutch.

It can be seen that such integrated farms are profitable and attractive commercial projects. Such farms open up marginal and unusable lands for development for the communities.

Another concept of using Aqua – Agro Complex along with power generation and Carbon Sequestration Process is given in Fig 2

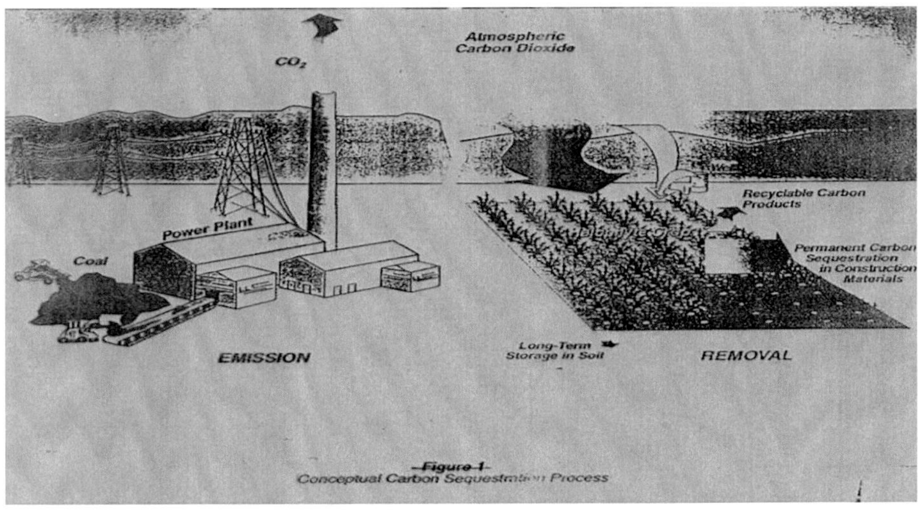

Fig. 2: Aqua Agro Complex

ENVIRONMENTAL EFFECTS

Massive plantations of Halophytes should have several environmental benefits.

Vast tree plantations are often suggested to absorb CO_2 and increase carbon storage in biomass. It is estimated that nearly 2 million KM^2 of new tree plantations would be needed to sequester just one third of annual carbon emission from fossil fuel burning. Sea weeds and halophytes are comparable to tree plantations in productivity (nearly 5tc/ha/yr). Sea weeds are relatively costlier to produce in the ocean sea farms as against halophytes.

The organic soil which would develop by flowing in halophyte biomass over a period of years may be used in future for conventional crops also. Massive halophytes plantation will improve the saline soil health and help fix atmospheric CO_2 in a big way.

There is, however, a caution. Monoculture has to be avoided and the ground water condition in undrained basins have to be watered.

Lately, looking at the overall situation of farmers, I have come to believe that rather than a large agua agro complex as was suggested earlier, we should encourage several small scale integrated units which can be managed by small coastal farmers / fishermen. A demonstration project was built at Ubhrat – Navsari later with integrated Aqua – Agro unit of small size which can be managed by a family.

CONCLUSION

Developments in Saline Water Agriculture will form an important and integral part of eco agri revolution. As demonstrated in a number of situations, saline water agriculture offers a new chapter for our agriculture future and offers much hope for new crops for food, resource utilization and climate change challenge.

What is more, the saline water cultivation of halophytes and integrating as aqua – agro complexes offers the great potential of Reversing the Flow of resource from 'Sea to Land'.

REFERENCES

1. Hodges C.N. and Kaufman A.M., "Solution from the Sea: Direct Sea Water Irrigation of Agronomic Crops ; ECO – Inform '96; Lake Buera Vista, Florida, 4-7 Nov. 1996.
2. Van Oosten H.J. and de will J.G. Bio Production and eco system development in Saline condition – N.C.A.R – The Haque, Netherland, nr 2000/10
3. Mehta M.H. Garg SK and Datta R.L. 'H-D Process of Desalination of Sea Water' – First National Conference on Desalination, CSMCRI – Bhavnagar, Nov. 1967.
4. Gupta I.C and Gupta S.K. 'Use of Saline Water in Agriculture, 3rd Edition, Science Publication (1) Jodhpur, 2003.
5. Reddy MP and Iyengar ERR, 'Crop Resources to Salt Stress, Sea Water Application' – Handbook of Plant & Crop Stress, Marked Dekkar, New York, 1999.
6. Boyko. H., Salinity and Aridity, The Hague, Junk, 1966.
7. Mehta M.H. Climate Change, Bio Shield – Invited Lecture, Cornell University, USA, March 2013
8. Mehta Purvi, Ed; Coastal Land Development Strategies – GSFC Science Foundation Meet, March 25, 2002
9. Bruno Ladeiro; Saline Agriculture in the 21st Century; Journal of Botany, Vol. 2012 (2012).
10. Personal Discussions with Carl N. Hodges – 1967 to 2011.
11. NAU Agricultural University, "Aqua – Agro Experimental Farm – Ubhrat", NAU-Report, August 2008.

Recall the face of the poorest and the weakest, and ask yourself, if the step that you contemplate is going to be of any use to him/her. Will he/she gain anything? In other words, will it lead to Swaraj for the hungry and spiritually starving millions.

— Mahatma Gandhi

18

ICT and Reaching Out with Eco Agri Model

All agriculture universities have three main Directorates – Education, Research and Extension. Of the three, I have often observed that Extension Department comes in for maximum criticism. 'Loose links'and 'need to do much more in transfer of knowledge and technology' are frequently emphasized at various levels – especially in meetings with farmers groups.

This is understandable – as it is said that less than 10% - often less than 6%, of lab scale knowledge and development reach the farmers fields in developing countries and even in many 'developed' counties. The situation is much more challenging when we consider a large population of small and marginal farmers – who as we saw can play extremely important role in eco-agriculture. A large percent of such farmers – women and men are practicing eco friendlyagriculture without realizing it. Their 'ignorance' can be converted in to advantage and new opportunity. The age old extension practice of reaching farmers through meetings, visit , seminars, camps, demonstrations all have limitations and need to be 'scaled up'. New developments in information and communication Technology offer a great opportunity for a big jump in reaching the unreached. Again to alarge population of 'Green Revolution' – there is a new hope and opportunity.

ICT IN AGRICULTURE

The potential for using ICT in agriculture can be divided broadly under two heads (1) as a tool for direct contribution to agricultural productivity and (2) as an empowering tool for the farmer to take and quality decisions.

There have been several excellent examples of practical use of ICT in Agriculture and expectations are rising such that a major transformation should take place in all aspects of farming and agribusiness. Some typical good examples are here:

E-KRISHI KIRAN PROGRAM (SOIL HEALTH CARD PROJECT)

Brief description of the programme/project/Initiative

Architecture of e-krishi kiran programme (Soil Health Card Project) The architecture of the e-krishi kiran programme consists of four components viz., (1) Input, (2) Data base and Software, (3) Network and (4) Output. The details are as under. (A) Input To achieve the objectives of the programme, utilizing various applications, a number of basic information are required to be fed to the system. These information / data were collected from various government departments, agricultural universities, and NGOs in the state. Some of the important ones are given below. (1) Information on individual farmer (2) Soil-test values of individual farmerâ€™s field: values of EC, pH, organic carbon and available phosphorus and potassium. (3) Soil characteristics (4) Meteorological data (5) Agro-climatic zone wise cultivation practices of all crops (6) Data on economics (Taluka-wise and crop-wise) (7) Recommended fertilizer dose (8) Information for crop planning (9) Agricultural Production Planning

List of the 5 achievements of the programme/project

1. The soil analysis of 32 lakhs individual farmers has been made and it •™s results are stored in the database. On that basis Soil Health Card has been generated and given to the farmers. Total 3183875 Soil Helth Cards are Generated.
2. The village Soil Health card based on fertility index from soil test values of the village fields are generated for 18,600 villages of the state, which is useful for farmers who•™s fields are not analyzed.
3. For all the 225 talukas of Gujarat, e-krishi kiran system generates alternative crop Planning & recommendations.
4. For all the 225 talukas of Gujarat, e-krishi kiran system generates alternative crop Planning & recommendations.
5. For all the 18600 villages of Gujarat, e-krishi kiran system creates Village Model Action Plan.

BOX – 1

WEDNESDAY, JUNE 18, 2014

Govt plans 'soil health cards' for all farmers

Vishwa Mohan TNN

New Delhi: Seeking to spur agricultural growth, the Centre will soon launch a comprehensive plan to provide 'Soil Health Card' to all farmers across the country.

The card will carry crop-wise recommendations of nutrients/fertilizers required for farms, making it possible for farmers to improve productivity by wisely using inputs. A computerised system will be developed allowing local agriculture science centres to keep details of 'soil test' results. Soil samples will be collected even from small farms in remote villages.

The system will eventually allow farmers to download the health card using 'unique number' allotted to each soil sample. This way any change in ownership of the particular farm land will not create any problem in getting such cards or getting it updated.

Though a few states including Gujarat, AP, Karnataka and Haryana had successfully distributed such cards a couple of years ago, most states did not make it operational beyond villages around various agricultural universities. The Centre's fresh move will make it universal. "Providing 'Soil Health Card' to all farmers across the country is our priority. We are working out the details, studying best practices being followed by different states," agriculture minister Radha Mohan Singh said. He said the move will help farmers in identifying "health of the soil" which will go long way in improving productivity through judicious use of fertilisers and water.

TAPPING THE GRASSROOTS

- Card will carry crop-wise recommendations of nutrients/fertilisers required for farms to improve productivity
- Computerized system to be developed allowing agriculture science centres to keep details of soil test results
- System will eventually allow farmers to download card using 'unique number' allotted for each soil sample
- Centre may seek help from agriculture science students [illegible] their first degree with their work in sample collection

SATELLITE KRUSHI GOSTHI : FARMERS – SCIENTISTS LINK

In the year 2000, after several rounds of brain storming and exercises about improving the extension and technology transfer at Gujarat Agricultural University, I had personal discussion with the team of RESECO (Remote Sensing & Communication - Gandhinagar). This team of ISRO working on a TV studio at Gandhinagar (capital of Gujarat State) and with several rounds of discussions, we decided to establish GAU Satellite Krushi Gosthi (GAU SATKRU) with one – way video and two-way audio. It was realized that for a big university like GAU (area and spreadwise it was considered the largest in Asia!) with 4 main campuses and nearly 65 experimental stations and farms, a facility like this can be a quantum jump in direct communication between farmers and scientists.

Our enthusiasm was high and the two teams worked hard to meet the deadline of five months which I had set in consultations with Dr. K. Kasturirangan (Chairman – ISRO) and the local team. It looked difficult in the beginning with the age old bureaucratic system but with the hard work the facility was completed and inaugurated in record time. We named it as GAU Satellite Krushi Gosthi or GAU SATKRU!

GAU SATELLITE KRUSHI GOSTHI

GAU SATELLITE KRUSHI GOSTHI

The Gujarat Agricultural University (GAU) performs triple function of teaching, research and extension education. The research generates technologies that can be used by farmers and ultimate users. The present system of transfer of technology from SAU to extension functionaries of development departments of the state and in turn to the ultimate users consumes considerable time. The GAU has prepared a major plan to apply new modern tools like Satellite Linkages for agriculture sector. The **GAU Satellite Krushi Gosthi** for transfer of technology can reduce the time lag to a considerable extent. The system will help for larger area coverage as well as will considerably reduce the distortion in message transfer. Since the proposed network provides facility for two-way dialogue, it will help in providing solutions of queries on the spot. This will help to keep a live contact between the scientists of the university and the farmers.

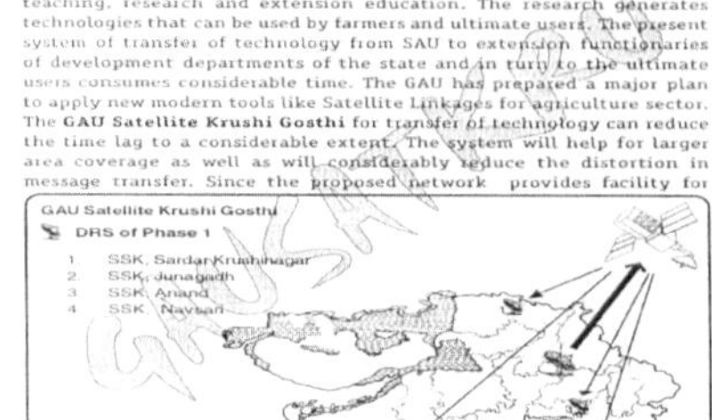

GUJARAT SATCOM NETWORK

A satellite based distance interactive education system normally consists of a number of remote classrooms spread over the state and a teaching end, which is centrally located. The Gujarat SATCOM Network has full capability for one-way video and two-way audio. RESECO has established SATCOM Network consisting of TV Studio at Gandhinagar. The video & audio from teaching end is digitally transmitted to the classroom end. The return audio is available through STD lines. At present, 78 DRS (Direct Reception Station) to receive transmission are already established throughout the state with different departments. It is expected to establish a total of 300 such DRS by various departments of Government of Gujarat by the end of year 2000, and GAUSATKRU will form vital linkages with all these.

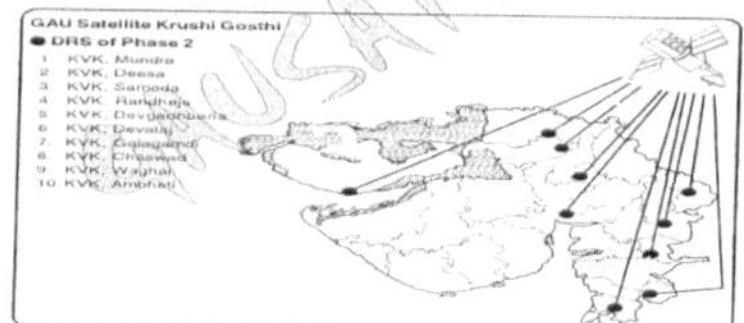

TRANSFER OF AGRICULTURAL TECHNOLOGY TO FARMERS

Initially, DRS at the four Sardar Smruti Kendras of Gujarat Agricultural University have been established. Necessary infrastructure is already developed along with purchase of equipments from own resources of the university. The system will enable to transmit farm technologies at a regular interval. With this, farmers will have greater accessibility to agricultural and related information. The farmers will be receiving technological know-how directly from the scientists of the university along with facility of interacting with them. The accessibility to farm information would encourage the farmers for application of new profitable technologies in their farming. It will also help them to receive information regarding inputs as well as markets. The system will also be used for communicating the latest information to students of the university, organizing lectures of dignitary/experts in agriculture, getting reflection from expert farmers and for interaction among the farmers groups.

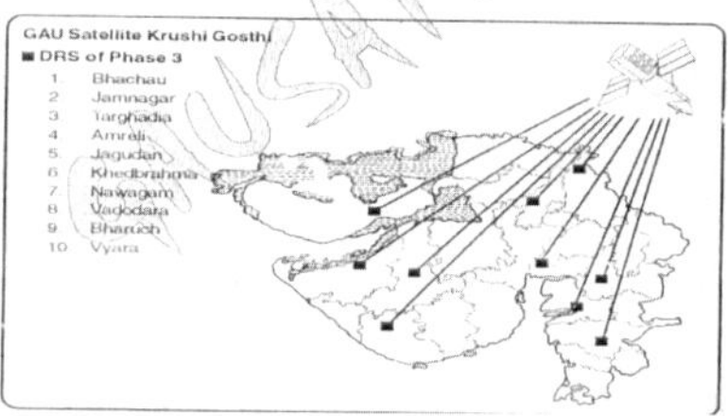

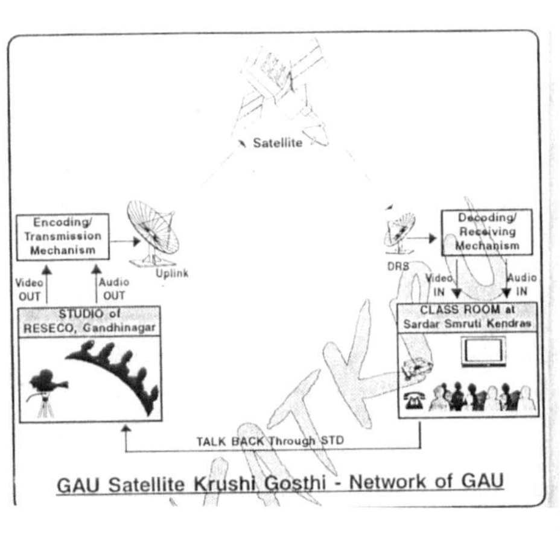

GAU Satellite Krushi Gosthi - Network of GAU

A JUMP IN COMMUNICATION AND TECHNOLOGY TRANSFER

The direct communication facility proved to be a boon for reaching even corners. As soon as the studio was inaugurated by the Governor of Gujarat in 2001, the direct dialogues brought new dimension. In fact, as seen in the picture the very first question was asked by a farmer woman – Bhaniben who lived in a small village near Junagadh(Saurashtra) and who had travelled in a tractor to participate in the GAU SATKRU program – more out of curiosity. I recall our conversation clearly even today -

GAU Satkru

Bhaniben – One question Sir!

I - Yes, please go ahead.

Bhaniben – Sir, it is very good that you have this new Antarix (space!) facility. I have two buffaloes – our main source of income – in what way will this help us?

(I think this was the hardest question I have ever answered – and that too with hundreds of farmers and scientists watching live in different parts of the State!)

After a few seconds I answered ...

I – "Bhaniben, for example – next week we have scheduled a program – 'animal nutrition and animal health' for the coming summer. Will this be useful to you?"

Bhaniben – Yes – That will be good. We will look forward to it. Thanks!

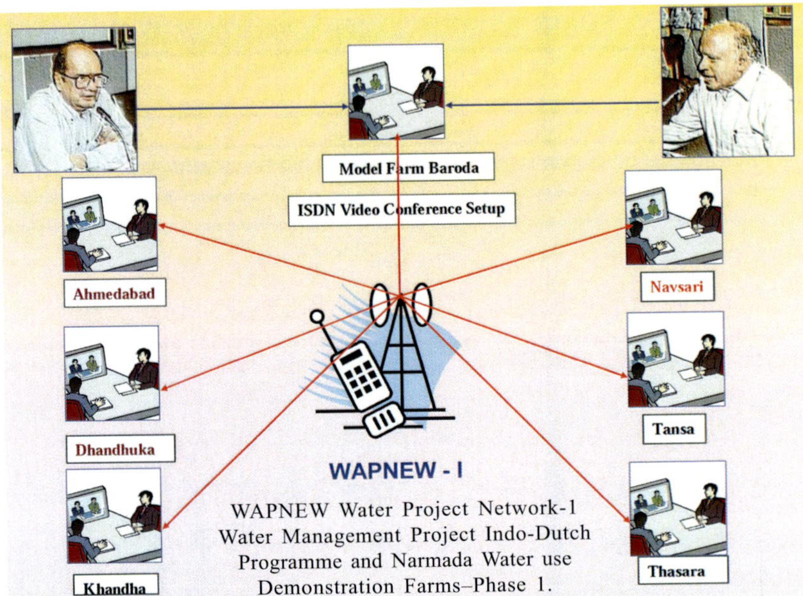

I have often discussed this conversation as an example of this technology as a great tool to reach people. Later, the same facility was used to prepare water management plans at different field stations in Narmada Canal areas – as shown in the figure.

Linking up with marketing yards using GAU SATKRU was another initiative that soon followed.

Over the period of time this experience became a valuable base for appreciating the role of ICT for extension and introducing newer initiatives in the state Gujarat and elsewhere.

Connecting with market yards

M KRISHI ICT PLATFORM FOR 20:20 MODEL

m-krishi was developed as an award winning innovative platform to connect farmers with a variety of stakeholders packaging multiple services through mobile phones by TCS (Tata Consultancy Services). It was introduced commercially in Panjab and UP and started with about 1000 farmers in the first stage and was a great success. Later while discussing at several forums it was decided to form a consortia for the 20–20 model.

A proposal was worked out for a 20:20 model application for 3 to 4 lead crops like Potato, Castor, Groundnut, Vegetables and fruits for the north Gujarat region. The partners of the consortia would be TCS (Tata Consultancy Services), TSA (The Science Ashram), SOFILWM (Society for Integrated Land & Water Management), NGAPCL (North Gujarat Agricultural Producer Co. and SDAU (Sardar Krushi Dantiwada Agricultural University).

ACTIVITIES

1. Extension Support – Support farmers to adopt modern agricultural technologies for crop yield improvement and input cost management.
2. Demonstration and Soil health Improvement – Conduct demonstration of good agricultural practices (GAP) at farmers field and improve soil fertility through proper application of fertilizers.
3. Capacity Building – Make farmers aware to use organic and chemical fertilizers & pesticides judiciously. Also make them aware on use of ICT based precision farming and extension services.
4. Availability – Make source of bio-pesticides and bio-fertilizers known to farmers if they are advised to apply them. This would be done by reaching agreements with suppliers to supply the bio-inputs at best prices to farmers through North Gujarat Agro Producer Company (NGAPCL).
5. Market Linkage – We will help NGAPCL to sign agreement with procurement agencies to purchase produce directly from farmers with better rates than other unorganized farmers.

Deliverables at the end of 3 years

- The gap between average productivity of lead crops/ activities generally obtained by farmers in the cluster and the yield obtained by progressive farmers would have been bridged by at least 50%.
- The average income of the farmers should go up at least by 30% in nominal terms.

- All farmers should be provided with credit facilities.
- Repayment of the bank credit should be at least 95% of demand.
- All farmers of the cluster should be part of farmer's collectives.
- A Business Collective like Society/ Producer Company etc., should be in place.

Such an ICT platform would enable personalization of services (mKRISHI or mobile KRISHI) and farmers stakeholders such a scientists at Universities, agriculture experts, agri input companies, rural banks etc can be "connected to farmers or farmers' organization.

BOX – 2

How Big Data is Revolutionizing Farming and Supply Chains

Agriculture is experiencing a revolution with the integration of decisions and tools driven by big data. Subsistence farmers need to close the yield gap. Production farmers want yield boosts and cost savings. Consumers are demanding healthier food. Data is being used in agriculture worldwide to increase efficiency, while lowering the environmental impact. Modern computing power has led to breakthroughs in the ability to collect, exchange, process, and synthesize data in ways that impact the use of farm equipment, seed optimization, fertilizer and crop inputs, irrigation, and farm management. In order to extract value from big data, it must be processed and analyzed in a timely manner and the results need to be available to make decisions that benefit farm operations. The effectiveness also relies on having the right combination of people, process integration and technology. According to the AgFunder Investing Report, over 500 food & agriculture technology startups raised $ 4.6 billion in 2015. The big data revolution is in its early days and it is difficult to know what it will look like even in two years. We are currently in the early adopter stage – with great one-off stories of wild success alongwith grumbling at the failures of beta tests of new products. Machine learning products and services will get exponential better with more users. Vonnie Estes, an Independent Consultant based in California, has the story for New Ag International.

A 20-20 Model, which helps reduce agri input cost by 20% or more and at the same time improves the farm productivity by more than 20% is demonstrated in different parts of the world for a variety of crops. Together with Soil Health Cards, checking the balance of Organic Carbon in the soil and other important elements like N, P, K, S *etc.*, it is possible to get a major jump in the farm productivity especially for the resource poor farmers.

How to take such a 20-20 Model to the remotest farms? ICT can play a major role here. ICT is the powerful tool to reach maximum number of farmers. Information on new innovations and development in the area of agri bio inputs can be effectively disseminated by the help of various ICT tools, which will ultimately improve the quality of farmers life at the grass root. The future extension mechanism will largely depend on ICT for providing appropriate and site specific technology for the farmers. It will support the large network of transfer of technology and in furnishing timely and proficient advice to the farmers. Some examples of application of 20-20 Model in different areas and crops in the world are established by the joint efforts of Gujarat Life Sciences, The Science Ashram (an NGO) and Agri Universities. These can be used for site specifications, application of the model for the farmers in different areas.

Mobile phones are currently the most widely used ICT among farmers. From farmers perspective this means of communication is one that they can control more readily than others, and they can use it as and when they like. Many farmers are picking up well – versed with their mobile for meeting diverse information and communication needs.

SRI (System Rise Intensification)

SRI – System Rise Intensification is a great Agro Ecological or Eco Agricultural Innovation. It derives from decades of observation, experimentation and collaboration with farmers by a French Priest, Fr. Henri de Laulanil, Rather than relying on improved varieties and purchased agrochemical inputs - (the strategy that succeeded in Green Revolution), SRI achieves its higher yield and other benefits with neither new seeds nor chemical fertilizers.

As stated by Uphoff "SRI succeeds just by changing the farmers manage their soil, water and nutrients. This makes the innovation very relevant and welcome for small and marginal farmers. With SRI, the greater yield is achieved with fewer inputs, since the methodology reduces plant populations and seed requirements by 80 – 90% and requirement of irrigation water by 25 – 50% and substitutes composts or other organic inputs as much as possible for chemical fertilizer. By skillful and appropriate use of their resources farmers get more output with less expenditure, and even with less labour".

The spread of SRI around the world in the initial years depended upon the use of email communication, backed up by a website on the internet. Now in recent years, a number of other ICTS have been utilized to bring knowledge and opportunities to millions of farmers to support their economic and social advancement.

SRI AND ICT

Norman Uphoff has been engaged in a considerable range of ICT for the world wide promotion of an unprecedented eco agri revolution. This experience, he has often stated, has engendered a confluence of theoretical and practical perspectives on empowerment of farmers through ICT. In fact, he and his teams have been engaged in more than 50 countries to reach and empower millions of small holding farmers around the world with ICT facilitating the promotion of changes in agricultural production methods, helping many householders to enhance their food security and raise their incomes, promote greater self-confidence and innovation, contribute to environmental sustainability, and mitigate climate – change efforts.

Information and communication is important factor of agricultural productivity enhancement.

ICT CONTRIBUTION TO SRI DISSEMINATION

Though SRI work and demonstrated yield improvement (8t/ha against earlier paddy yield or 2 t /ha) was done in the yields of small farmers in Madagaskar, (The ICT support for SRI dissemination started later. The special SRI – website (http://Sri.ciifad.cornell.edu) has grown and became a major free access source of knowledge and communication. Later the SRI International Network and Resources Center (SRI – Rice) was established and the website has thousands of visitors from more than 100 countries. It was also joined by many other websites like SRI – India, WASSAN (Water Shed Support Services) etc.

The development and impact of ICT for SRI are still 'a work in progress'. It has clearly emerged as a great examples of empowering people – particularly the small farmers and changing the knowledge, skills to a new level.

The above typical examples are only indicative. There are several such path breaking initiatives like MSSRF - Virtual University, ITC Chaupal, Digital India, Warna Cooperative, Amul etc. which show the increasing awareness and applications. Agriculture will experience a huge change with ICT and small and marginal farmers should particularly gain.

CONCLUSION

ICT revolution is unfolding and has high visibility. In agriculture extension and technology transfer it will be a unique tool. Especially for the small and marginal farmers it can provide an unprecedented help to join the eco agri revolution.

REFERENCES

1. MSSRF (2009). "Reaching the Unreached – Community based Resource Centers, MSSRF – Chennai, 2009.
2. Mehta M.H. (2014). Taking to the Farmer – Farmer to Farmer Initiative for Eco Agriculture, Borlaugh Centenary Dialogue, MSSRF, Chennai, March 13 – 15, 2014.
3. Shekh AM (2011). e-Krishi Kiran (Social Health Card) Program, AAU – Report, 2011.
4. Vishavjeet Chaudhary and Gursharan Singh (2016). "What is the future of Agriculture in India, Wire, 19.7.2016.
5. Mittal S.C, Role of IT in Agriculture in India, IFFCO; Parts Semantic Scholar.
6. Mehta M.H. (2010). 'Reaching Out through ICT for the 20-20 Program' Invited talk at 7th Convention on Grameen Gyan Abhiyan (Rural Knowledge Movement), Mumbai, 27 – 28 November, 2010.
7. Personal Conference – Dr. Pande and Dinesh Kumar, m-KRISHI – Tata Consultancy Services.
8. Norman Uphoff; Empowerment of Farmers through ICT; ESOSOC Expert Group Meeting – UN HQ, NYC, 10-12 September, 2012.

"Vaho Vishwamitri" – (Flow Vishwamitri Flow)

— School Children's Song - Baroda

19

Model River Revival System – Vaho Vishwamitri Abhiyan

Rivers are the most important system for water, for irrigation and human settlement. Historically, civilizations always nucleated near rivers, and latter also had social and religious significance. Rivers in India are considered holy and play the most important part in the lives of people, fauna & flora and the future environmental system.

The Ganga Basin is the largest river basin in India which includes the Ganga, the Yamuna, the Ghagra, the Gandak, the Kaveri basin etc. Compared to these, the Vishwamitri River, situated between Mahi River in North and Narmada River in South in Central Gujarat is a small seasonal river but it is very important for the city of Vadodara and Central Gujarat region. This seasonal river originates from Pavagadh hills and flows east to west. It is a typical small river with the kind of problems in most rivers in India and other countries – liquid and solid wastes, enouncements, blocking of water ways, floods in monsoon months and dry dumping basin during most other months etc.

RIVERS AND THEIR REVIVAL

River basin planning and watershed management were the centerpieces of water resource planning at the dawn of the era. For most of the century, however, natural resource policies and pressures led to watershed degradation. Today, the resources value of watershed ecosystems are being rediscovered. Watershed management reflects the recognition that we can only sustain biodiversity by managing the future ecosystem.

VISHWAMITRI RIVER

In a well researched book 'Vishwamitri Mahatmyam' edited by Thaker of Oriental Institute – Vadodara the history, glory and mythological details of the ancient river Vishwamitri has been brought out. It is said that the river was earlier known as Kaushiki and was later renamed as Vishwamitri in honor of the great Rishi Vishwamitra who did penance on its banks. The river originates from mountain 'Pancavakra' – now known as Pavagadh situated in the Panchmahal district of eastern Gujarat and flowing to the west. It passes through the city of Vadodara (Baroda) and meets the river Dhadhara which flow down the Gulf of Khambhata (can bay in Arabian Sea). The Mahatmya is a type of literature where in generally the significance and importance of the river and religions places are narrated with anecdotes.

The whole work is written in a dialogue form in the fashion of the Purana texts. It also narrates that the entire region was known as forest 'Sankara' and gives a list of vast verities of animals, trees, shrubs, grasses and birds.

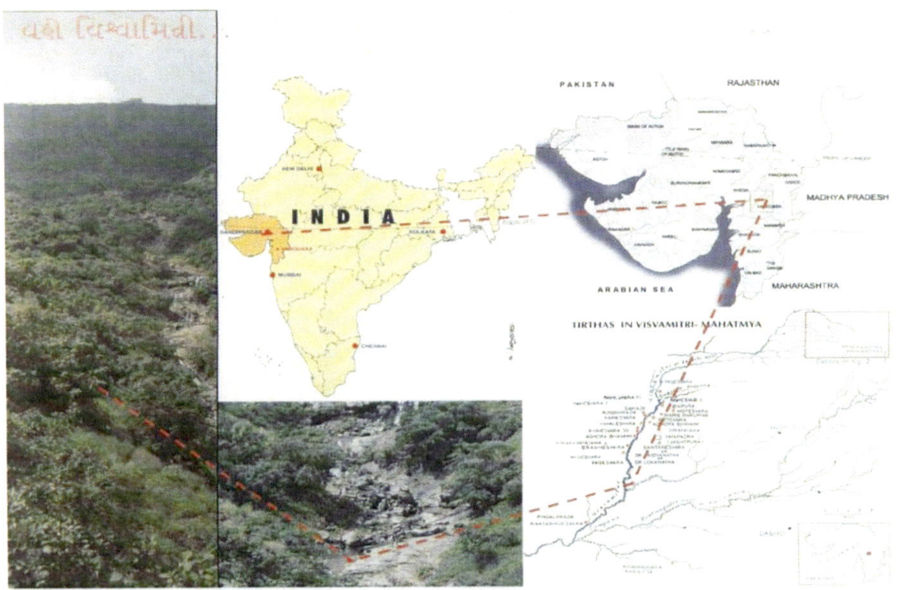

Vishwamitri River – Central Gujarat

VAHO VISHWAMITRI ABHIYAN – PEOPLE'S MOVEMENT

Over the period of time, this once beautiful river became a waste dumping ground, encroachment and misguided urban growth causing floods, water logging in monsoon and waste water drain. This is depicted in photo/figure.

A movement – Vaho Vishwamitri Abhiyan was started to revive this ancient – once beautiful river. The focus is to create a nature system that needs few human interventions for self sustaining river and healthy ecosystem with adequate green cover to sustain life. It brought a multidisciplinary team of experts in various fields like civil & other engineering, architects, geologists, botanists *etc. etc.* on a common platform. Later hundreds of citizens, representative from Government departments, school children, farmers and citizens from urban and rural areas joined the movement.

In working out a 'Project Proposal' for a holistic proposal the basis were in three zones as described in the Annexure I.

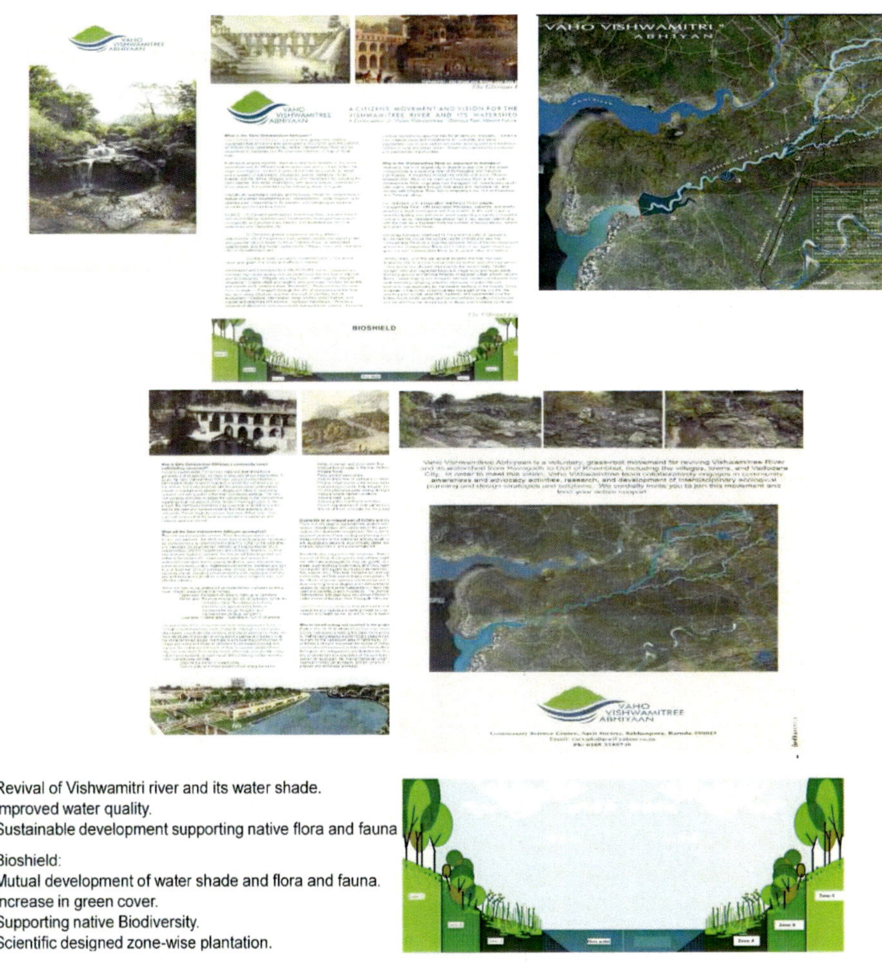

Annexure 1

Contd. Annexure 1

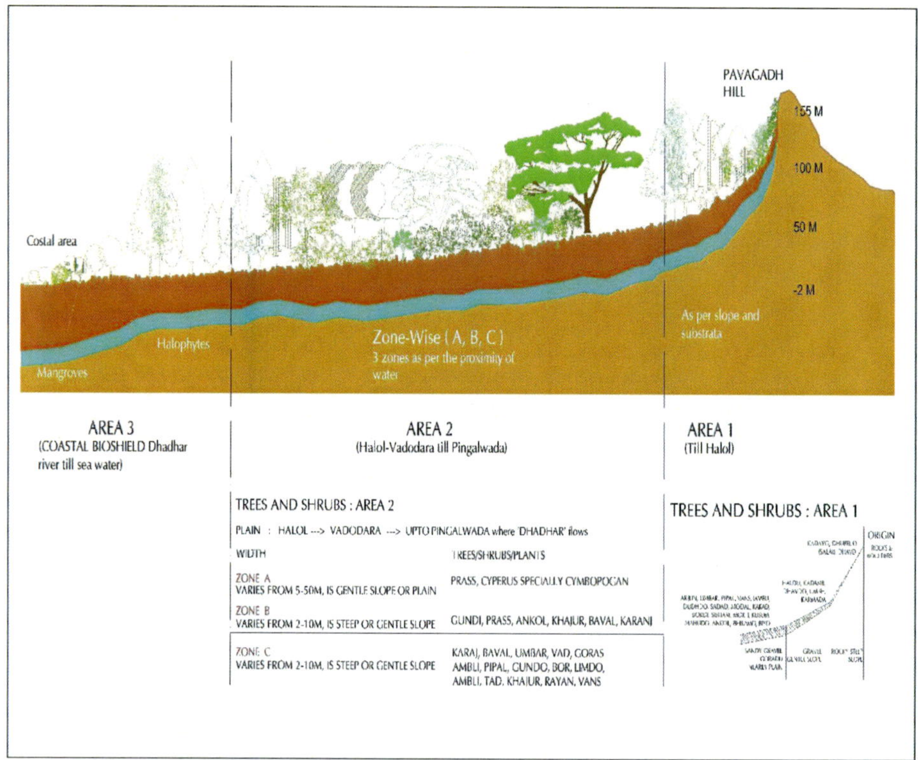

In the local language (Gujarati) the word Vaho connotes 'flow'. The river is 'urged' to flow in order to revive. With the support data from Bisag (Bhaskaracharya Institute for Space Application Gandhinagar) and ISRO, exercises were taken up regarding Check Dams (small) and water bodies. The preliminary calculations showed that with about 7 check dams all along the length of river. The amount of water that can be stored – even if only 15% rain water is captured – will be 150% more than the present main water supply Lake Ajwa for the city as shown. Most of such water during monsoon creates flooding problem!

BIO SHIELD

The second important component is Bioshield. The entire river basin – nearly 135 km from Pavagadh to Gulf of Cambay is classified in three areas as shown in the figure.

The design of scientific plantation – Bioshield also demands the consideration for water regime and availability of plant species to specific micro-environments. Hence – as shown these areas have been planned in further vegetation zones transversely as zones A, B and C. Each tree and shrub acts as a micro check dam and a careful planning can be the most important component of a river revival programme. An exercise was taken up by experts in Botany, Agriculture, Forestry and Landscape Designers to prepare zone wise list of plants and shrubs which is listed as Annexure to this chapter.

It was worked out that nearly 90 million trees and shrubs will have to planted.

Bioshield - concestual

SEED BALL CAMPS

Since the entire plan was very convincing to people, more and more people started joining the movement. Tree plantation camps were organized at different locations on the banks of the river. The support of forest department and other bodies readily came. However it was also realized that for such massive plantations newer technology development became necessary.

'Seed ball' or 'Seed Palleting' technology development was the important requirement. This is partly discussed in the chapter on Bioshield. Earlier work was taken up at the labs of The Science Ashram/Gujarat life science to develop ecofriendly coating mixture for seed balls - made of bio-fertilizers, bio-fungicide, growth enhancer and a bio polymer as a binding agent.

BOX – 1
Seed Ball / Pallet Technology

Seed ball or Seed pellet Technology for

VAHO VISHVAMITRI ABHIYAAN

What is a 'seed ball'?

"Seed ball" or "Seed pellet" is an innovation that has the potential to revolutionize the Indian farming industry. It is a brilliant idea that, involves coating a regular seed, with a mixture of organic growth enhancers, bio- fungicides and a special polymer that acts as a binding agent.

Why 'seed balls'?

The coating avails the seed an even shape with increase in size and weight, which ensures greater planting and dispersion efficiency. The coating is also enabling a free flow of oxygen and moisture retention to the seed, to ensure germination. Seed ball technology does not need seed to be sown manually. Seed balls can simply be tossed, making them very useful for farming on hills, large plain areas, ravines, river banks etc.

Importance of seed ball in the Vaho Vishvamitri Abhiyaan

Vaho Vishvamitri Abhiyaan is for the revival of Vishwamitri river water. The Bioshield designed along the length of the river require the plantation of more than 9 million shrubs and trees of different species supportive of native biodiversity. The seed ball technology will be useful to develop bioshield not only on Pavagadh hill but on river bank areas that are difficult to approach so also so, the saline areas before the river meet the sea. The seed ball coating mixture has been scientifically developed and prepared at the Science Aashram and Gujarat Life Sciences laboratory. Vaho Vishvamitri Abhiyaan organized by the Community Science Centre Vadodara in association with the other organizations, will take up the process of making the seed balls from the mixture and eventually also execute the plantation of these seed balls along the Vishvamitri river involving students of various schools, colleges, and interested individuals and elite citizens.

(Nirja Bhatt & Dr. M.H. Mehta)

Camps were organized at different locations and especially difficult to reach areas like hills, ravines, depressions and areas where crocodiles are present! Children especially enjoyed making seed balls and then going to sites and throwing them while chanting 'Vaho Vishwamitri' chorus! Some attempts were also made with modified 'drones' as the carrier of seed balls & more innovations are expected in this line!

Seed Ball Camps

WASTE MANAGEMENT AND ORGANIC FARMING

Use of ecofriendly waste water treatment for augmentation of water supply, segregating solid organic waste and converting to biocompost and eco agriculture/ organic farming was the third important component of the holistic plan. Several wastewater treatment plants along the length of the river can separate out & recycle water after treatment and the sludge with treatment and conversion to biocompost is proposed for cities. On the other hand – the Clean Village – Eco Village proposals discussed elsewhere in this book are worked out at village level where small cooperatives can be formed to set up village level biocomposts plants.

Coinciding with the 'Organic Farming Policy of Gujarat Government – a proposal for organic farming on the both the banks of the river up to 5 km was prepared. It is well-known that around water bodies and rivers banks the use of chemical fertilizer and pesticides is higher and the runoff from the farms cause much pollution. A model of integrating waste management, biocomposting and organic farms on the bank was therefore welcomed by the state government who conveyed that such an initiative can even create a brand 'Vishwamitri organic food' for the region. Such a plan will need involving farmer bodies and villages and can be ideally adopted elsewhere. Mr. Rajendra Singh – the well-known water man of India – visited the Vaho Vishwamitri Abhiyan camps and round-

'Vishwamitri's revival can be a model for rivers in India'

Times News Network

Vadodara: Revival of the ancient Vishwamitri river can become a model for revival of other rivers in the country that are facing similar problems of pollution and destruction, R Chidambaram, principal scientific advisor to Prime Minister Narendra Modi said here on Friday.

Chidambaram was in the city as chief guest at the national round table on 'Model river revival system - Vaho Vishwamitri Abhiyaan'. The abhiyaan is a people's movement that brings together engineers, environmentalists, school students, teachers, government and the society at large.

MY CITY MY RIVER

According to Chidambaram, the movement is country's first and biggest unique model where all stakeholders of the society come together to revive a river.

"The best part of the abhiyaan is the active involvement of children and school students. Rivers across the world are facing degradation due to human encroachment, unregulated urban growth and ignorance. Therefore the case of the revival of Vishwamitri can also be looked at from the global context," said Chidambaram.

Experts say people's movement is a unique mission to revitalize river

Last year, the National Bioshield Society (NBS) and the Community Science Centre (CSC), Vadodara had submitted the project proposal to the Ministry of Earth Sciences of Government of India.

"Over the last several decades, Vishwamitri river has become a dirty and neglected backyard of the city. The 'Vaho Vishwamitri Abhiyaan' is a call to revive the river by implementing sensitive eco - engineering, landscape planning and architectural designing to achieve the outcome," said former vice-chancellor of Gujarat Agriculture University Padmashri awardee M H Mehta, who heads both NBS and CSC Vadodara.

"As part of the river revival system, we have proposed to build check dams and linkages of other water bodies with the river, create a bioshield by planting 60 lakh trees, conduct organic farming and build gardens on both banks of the river and implement water and solid waste treatment using advanced techniques to clean the water," added Mehta.

The plan includes setting up seven lakes along the river so that city gets water enough for a population of 18 lakh and double than what it gets from Ajwa by 2025.

"This project is of prime importance for Vadodara. By changing the quality of water, you can also change quality of life of the people," said professor Erico Drioli, an international expert of water treatment.

table meetings at Vadodara several times and felt that such a model can be adopted around the 7 Rajasthan rivers which he has already helped in reviving.

A MODEL FOR RIVER REVIVAL

A holistic model like Vaho Vishwamitri is the right and sustainable way for sustainable revival of a river system. The modelonce prepared was presented and discussed at several forums and got much support from experts, leaders and people. The former President Dr. APJ Abdul Kalam took special interest and released the 'project proposal' and declared Vaho Vishwamitri as a great and doable project.

Later at a National Round Table on Vaho Vishwamitri experts from different parts of India & overseas participated enthusiastically and Dr.Chidambaram – Principal Scientific Advisor – GOI; even called this as a Model Project for all rivers!

It was also discussed by people involved in larger projects like Ganga Abhiyan and Yamuna BachaoAbhiyan that the best way for river clean up would be implementing such holistic plan for the tributaries of such big rivers.

SUNDAY TIMES OF INDIA, AHMEDABAD
AUGUST 11, 2013

TIMES CITY

Kalam's support to Vishwamitri's revival

People's President Makes Students Pledge For Clean, Green And Beautiful River

TIMES NEWS NETWORK

Dr A P J Abdul Kalam released project proposal of 'Vaho Vishwamitri Abhiyaan' during a function held in city on Saturday; (R) Kalam on Saturday addressed International Youth Conference organized by Ramkrishna Mission Vivekananda Kendra at Professor C C Mehta Auditorium inside M S University campus

Vadodara: The call to revive Vishwamitri river has received support from former President Dr A P J Abdul Kalam. Hailed as the people's President, Kalam on Saturday motivated and asked students to take an oath to clean up the water body and rejuvenate the green surroundings of the river that passes through the heart of Vadodara. In his typical style of asking students to repeat after him, Kalam said "Vishwamitri is possible".

He released the project proposal of 'Vaho Vishwamitri Abhiyaan' prepared by the National Bioshield Society and the Community Science Centre, Vadodara during a function at [illegible] School in Sama.

"If students think they can do it, it will happen," said the Missile Man of India, as he asked the students to repeat after him. "I can do it, we can do it, and India will do it. Vishwamitri is possible."

Praising the community call to bring life back into the ancient Vishwamitri river, Kalam said it is a voluntary grassroots movement for reviving the river and its watershed from Pavagadh to Gulf of Khambhat including villages, towns and Vadodara city.

"It is indeed a big mission. It needs big thinking. Of course, it needs big action. I am sure your mission will succeed with bioshield and eco-friendly sustainable system," he said.

Kalam motivated the group of students, teachers and eminent citizens who were part of the gathering to work on this ambitious project by citing examples of how individuals and groups have changed and brought life back to water bodies across the country.

Citing example of Sant Balbir Singh Seechewal in Punjab, Kalam said he has visited Sultanpur Lodhi twice. The Kali Bein rivulet, the place where Gurunanak Dev is said to have received enlightenment and which had over the centuries turned literally into a sewage ridden, weed choked drain, is today flowing clean and proud.

He asked students to take the oath that each one of them will plant five trees in the Vishwamitri region and nurture them. Citing contribution of a single individual from Gujarat – Premjibhai – whom he referred as a man with a tree mission, Kalam said the man who had begun his mission in 1987 by setting out every morning with a bagful of seeds and small spade in hand, is today responsible for planting more than ten million trees.

Later in the day Kalam also addressed International Youth Convention organised by the Ramkrishna Mission Vivekananda Kendra at C C Mehta Auditorium inside M S University campus. In the inaugural session of the two-day youth convention, Arunima Sinha, the first woman amputee to conquer Mount Everest also remained present.

Though, in principle all Governments – State as well as Central – are keen about a river revival project, often the tendency is for short term gain or even implementing 'cosmetic' solutions. In many big cities, river front projects are implemented by building cement concrete structures on a few kilometers of the river in the city area. This is neither desirable nor a sustainable way of river revival. It was also realized that when a holistic model like above is prepared and presented the understanding and enthusiasm of people goes up. That also brings about as great change in the understanding and attitude of Government authorities and bureaucracy who would otherwise adopt a short term and fractional solution.

There are encouraging signs now of acceptance of this holistic modules and beginning of the implementation program. In the holistic programme, the river front through the city will become a 'part' and not 'only' plan for the river!

CONCLUSIONS

Rivers are the integral and important part of our environment and agriculture system. Their holistic and sustainable plan is of paramount importance in eco-agricultural revolution. It is a doable and desirable way. When it becomes 'everyone's dream and everyone's mission, there are visible signs for a better tomorrow.

REFERENCES

'Vishwamitri mahatmyam' – Ed. JayantThaker, Oriental Institute, Gaekwad's Oriental Institute, Vadodara (India), 1997 Series No.176.

Vaho Vishwamitri Abhiyan – Brochure – Community Science Center – Vadodara – 2013.

Vishvam Pancholi; Analysis of runoff for Vishwamitri River watershed using SCSCN method geographic information system. Nat.Conf.are Transportation & Water Resource Engineering NCTWE – 2015.

Mehta M.H., Vaho Vishwamitri Abhiyan, invited talk presentation: Water Unlimited Seminar, Rotary Int., Mumbai 28.1.2012.

A.P.J.Abdul Kalam, Mission of Revival of Vishwamitri Abhiyan, Address to students participants in the Vaho Vishwamitri Abhiyan, 10th August 2013.

Vaho Vishwamitri Abhiyan, Annual Report by Community Science Center – Vadodara – 2008.

Bioshield Model for River Revival. National Bioshield Society Project Proposal for Mini. of Earth Science, GOI, August 16 2013.

R.Chidambaram – Inaugural Address National Round Table on Model Revival system – Vaho Vishwamitri Abhiyan CSC, Vadodara, 20th March 2015.

Ela Production's Nadi Ki Kahani Bundo Ki Jubani, Short film on Vaho Vishwamitri Abhiyan.

Annexure - 3

Vaho Vishwamitri Abhiyaan has following plan for its plantation
Zone I: Pavagadh hill, its valley and foot hill.
Zone II: Halol, Vadodara till saline affected area beyond its union with Dhadhar.
Zone III: Saline area to river mouth in Khambhat.

Design for Zone I: (A) Pavagadh hill top, (B) The River Valley and (C) Foot hill

Zone I (A) : Pavagadh hill Top

Micro habitat	Gujarati Name	Botanical name
Trees- rock crevices,	Khadak Payar	*Ficus arnottiana*
near water stream etc.	Kadayo	*Sterculia urens*
On boulders and rocks	Dhup gugal, Saledi	*Boswellia serrata*
Hill slopes: away from water	Kakad	*Garuga pinnata*
stream among rocky terrain	Modal	*Lannea coromandelica*
	Haldu	*Haldinia (Adina) cordifolia*
	Umbh	*Miliusa tomentosa*
	Biyo	*Pterocarpus marsupium*
	Kodaro, Khavas	*Firmiana colorata*
	Kothu	*Feronia elephantum*
Shrubs- rocky and gravel soil	Thor	*Euphorbia nivulia*
	Dhavdi	*Woodfordia fruticosa*
	Karamda	*Carisa congesta*

Zone I (B) : Hill Slope: The river valley

Micro habitat	Gujarati Name	Botanical name
Trees close to stream	Arjun	*Terminalia arjuna*
	Jal Jambu	*Syzygium heyneanum*
	Jambu	*Syzygium cumini*
	Karanj	*Pongamia pinnata*
	Umbar	*Ficus racemosa*
Away from stream	Sadad	*Terminalia crenulata*
	Dhavdo	*Anogeissus latifolia*
	Kothu	*Feronia elephantum*
	Khakharo	*Butea monosperma*
	Ragat rohido	*Tecomella undulate*
	Timru	*Diospyros melanoxylon*
	Kanajo	*Holoptelia integrifolia*
	Andrakh	*Anogeissus sericea*
	Mahudo	*Madhuca latifolia*
	Charoli	*Buchnania lanzan*
	Asan	*Bridelia retusa*
	Piplo	*Ficus religiosa*
	Vaas (Kantivas)	*Dendrocalamus strictus*
	Manvel vaas	*Bambusa arundinacea*
Liana/ Climber in-between	Dodi	*Leptadenia reticulata*
shrubs and trees near and away	Khir dodi	*Holostemma annularis*
from stream	Madhu nashini	*Gymnema sylvestre*

	Moti dodi, Malti	*Dregea volubilis*
	Dhamas	*Combratum ovalifolium*
Herbs, Grasses	Let it be natural hill grasses and herbs, need not disturb Ketki or *Agave americana* shall be on soil erosion site.	

Zone I (C) : Foothill

Micro habitat	Gujarati Name	Botanical name
Near stream	Arjun	*Terminalia arjuna*
	Jal Jambu	*Syzygium heyneanum*
	Jambu	*Syzygium cumini*
	Karanj	*Pongamia pinnata*
	Umbar	*Ficus racemosa*
Away from the stream	Vad	Ficus benghalensis
	Piplo	*Ficus religiosa*
	Mahudo	Madhuca latifolia
	Andrakh	*Anogeissus sericea*
	Charoli	*Buchnania lanzan*
	Asan	*Bridelia retusa*
	Sadad	*Terminalia crenulata*
	Vaas (Kantivas)	*Dendrocalamus strictus*

Zone II: Halol, Vadodara till saline affected area beyond its union with Dhadhar

Zone II (A) : Along river stretch close water body near habitation and agri. fields

Micro habitat	Gujarati Name	Botanical name
Trees close to river water (A)	Arjun	*Terminalia arjuna*
	Jambu	*Syzygium cumini*
	Karanj	*Pongamia pinnata*
	Baval	*Acacia Arabica indica*
	Prass	*Tamarix aphyla*
	Umbar	*Ficus racemosa*
Trees on bank slopes (B)	Amli	*Tamarindus indica*
	Bor	*Zizyphus jujube*
	Gunda	*Cordia dichotoma*
	Kantas Vaans	*Dendrocalamus strictus*
	Ambo	*Mangifera indica*
	Khajuri	*Phoenix sylvestris*
	Ravan Taad	*Hyphaene dichotoma*
	Rayan	*Manilkara hexandra*
	Baval	*Acacia indica*
	Karanj	*Pongamia pinnata*
	Markhor/ Dheki	*Diospyros cordifolia*
Away from bank water (C)	Kothu	*Feronia elephantum*
Nr. resi. area & agri. fields	Khakharo	*Butea monosperma*
	Ragat rohido	*Tecomella undulate*
	Timru	*Diospyros melanoxylon*
	Kanajo	*Holoptelia integrifolia*

	Andrakh	*Anogeissus sericea*
	Mahudo	*Madhuca latifolia*
	Piplo	*Ficus religiosa*
	Vaas (Kantivas)	*Dendrocalamus strictus*
	Vad	*Ficus benghalensis*
	Manvel vaas	*Bambusa arundinacea*
Among trees (shrubs on slopes)	Bordi	*Zizyphus mauritiana*
	Gundi	*Cordia gharaf*
	Karamda	*Carisa congesta*
Liana/ Climber in-between shrubs and trees	Dodi	*Leptadenia reticulata*
	Madhu nashini	*Gymnema sylvestre*
	Dhamas	*Combratum ovalifolium*
	Madhumalti	*Quisqualis indica*
	Shankhpushpi	*Clitoria ternata*
Herbs, Grasses	Natural grasses and herbs, need not disturb *Agave americana* shall be on soil erosion site	

Zone III: Saline area to river mouth in Khambhat (Pingalwada onwards...)

Plants including Mangroves, Halophytes & salt tolerant plants	
Herbs (A) close to water on mud flat	*Brachiaria* sp., *Sueda fruticosa* *Salicornia brachiata*
Mangroves (B) just near water, during high tide inundated	*Avicennia alba* *Avicennia marina* *Rhizophora mucronata* *Ceriops tagal*
Away from water (C)	*Salvadora persica* *Salvadora oleoides* *Casuarina equisetifolia* *Clerodendron pholomoides*
Economic plants (Herbs)	*Vetiver* grass, *Aloe vera, Ipomoea pes-carpae, Sporobolus marginatus Haloxylon salicornicum, Atriplex nummularia, Simmomdsia chinensis, Citrullus colocynthis*
Economic shrubs & trees Drought tolerant trees those can be planted inlands for coastal area	*Ricinus communis, Simarouba, Acacia* *Melia azadirachta, Acacia farnesiana, Acacia nilotica, Acacia Senegal, Acacia tortilis, Azadirachta* indica, *Prosopis cineraria, Pongamia pinnata, Cassia fistula, Eucalyptus teriticornis, Tamarix articulata Phoenix* sp., *Embelica officinalis, Punica granatum*
Aromatic Plants, Grasses and Forage	*Cymbopogon citrates, Catharanthus roseus, Plantago ovate, Matricaria recutita, Panicum antidotale, Panicum coloratum, Leptochloa fusca, Chloris gayana, Gatton panic, Hybrid Napier*

(Dr. Jitendra Gavali)

20

Remembering Dr. Kalam

The Late Dr. APJ Abdul Kalam is described as a great patriot, visionary leader and the 11th President of India (2002 – 2007). The country would have celebrated his 84th birth day on 15th October 2015. To many like me, he was the most ideal role model that the youth of India needed, in recent time. He emerged as a complete human-being who touched the lives of people.

My first meeting with him was when he visited IPCL in eighties - as the DRDO Chief to discuss about Carbon Fiber and Composites. I was heading GSFC's R&D Centre and other facilities like engineering plastic Nylon – 6 and their Composites and hence was invited to join the discussions. That interaction left a strong impression in my mind. Later when I became Vice Chancellor of Gujarat Agricultural University, I had many more opportunities to meet him and was often surprised at his deep rooted interest and knowledge in every sphere of Science & Technology. Subsequently, to my delight, I received an invitation from his office on the eve of 26th January evening - at home with the President and later was given more time for a one-to-one meeting the next day. Walking across the corridors of the Rashtrapathi Bhavan, the atmosphere looked rather formal and uneasy for me, but, as I was ushered into his room, I saw Dr. Kalam wearing his favourite blue shirt and smiling. He moved forward and greeted me with the same friendliness and warmth. As we started talking about Agriculture scenario, the work I was planning and the future prospects, I marveled at his understanding and insight and left the room truly inspired.

Later, I was fortunate to have many more meetings and interactions with him. In fact, during my frequent visits to Delhi as the Chairman of National Agri Technology – NRDC, I would call at his residence in the evenings at Rajaji Marg to discuss different developments and get his guidance. It soon became clear to his efficient Secretaries to inform me in advance and arrange a meeting with him whenever he would visit Anand or Vadodara. In one such meeting at the Circuit House in Vadodara, I took my 8 year old Grand Daughter alongwith me. He enquired briefly about the progress in my work in Bio-Technology and Agriculture in the first 2 minutes but later turned his entire attention to the little

girl. It was amazing the way he interacted and talked with much love and patience with the small child. That day, I saw another remarkable side of this great man. As we came out, my Grand daughter looked extremely delighted that she had found a new friend !

Eco Agriculture : Visit to The Science Ashram (TSA) and Gujarat Life Sciences (GLS) - Vadodara

In the winter of 2011, I got a message from his office that he would be visiting Vadodara for a school function and I could meet him in Circuit House some time. I suggested whether this time he would like to visit our small unit - The Science Ashram / GLS - working in the area of new generation Agri Bio products about which I had discussed with him earlier. His Secretary Prasad said, he would check and come back to me.The next day, I got a call saying that he

would visit for about half an hour in the afternoon. I informed Prasad that we are a small unit and perhaps not equipped for Securities and protocol requirements. I was assured that they would inform the concerned Govt. depts. (Collectorate, Police, etc.) that this was a personal and friendly visit and so the protocol should be minimum.

However, when he arrived at our facilities, there was a surge of security people around and I became a bit uneasy. Dr. Kalam just smiled and touched me on the back to convey that this was OK! Our young team of Scientists and Technicians were excited to see him taking keen interest, asking very relevant questions and giving encouraging tips. During a pause, he turned to me suddenly and asked - "Can I spend 2 hours here ?" I, of course, very happily agreed. In the end, he stayed for 2 hours and 45 minutes!. He went around the Research and Production facility for Bio Fertilizers, range of Bio Pesticides and Multi Microbial Technology for Waste Management etc. Soon after that we had organized his small interaction meeting with selected progressive farmers, Vice Chancellors, Scientists and press. It was a remarkable experience to watch him interact with them, talking at length about the Agriculture scenario, challenges, green revolution and the need for the new Agri Bio Revolution. On this occasion he had also agreed to release my book "Science, Spirituality and the World Tomorrow". Since the book was in Gujarati, he joked "I would like to carry one copy of this in Gujarati but promise me that you will translate it in English". I promised him, but alas! he passed away before I could complete the translation!

After the program and giving photographs with the excited audience, we went to my office for a cup of tea. As he entered the office and saw the photograph of Dr. Vikram Sarabhai on the wall, he smiled and said "Hey, you have kept the photo of my GURU here !" to which I replied, "my GURU too". He looked at one of my colleagues and murmured ….."GURUBHAIs".

While leaving my office, he put his hand on my shoulder and said, "my friend, this work on Eco Agriculture is so important and should be your mission". These simple words will remain with me forever as the guiding light. For everyone present there including our young Lab Assistants, Staff and Scientists - Dr. Kalam's visit has became a lasting memory to cherish.

Vaho Vishwamitri Abhiyan (VVA) – A model River Revival System

The river Vishwamitri, passing through the city of Vadodara, is an ancient and once a beautiful flowing river. It had a thickbioshield around it and clean flowing water. Like it has happened in case of many of the Rivers over time, now it too became dirty and stagnant. Through the Community Science Centre Vadodara and many other institutions we created a common platform involving experts from different fields, enlightened Men and Women and farmers in the region to take up Vaho Vishwamitri Abhiyan - the River Revival project. A very large number of school children also joined the movement.We prepared a holistic vision for the entire river starting from the hills of Pavagadh from where the River originates and then passes through cities like Halol, Vadodara, Bharuch etc. before meeting the Gulf of Cambay. An important component of such a plan was creation of Bio shield on both the banks of River. Earlier, we had implemented one bio shield project through a consortia of several NGOs lead by The Science Ashram (TSA), M.S. Swaminathan Research Foundation (MSSRF) and Indira Gandhi Centre for Atomic Research (IGCAR) at Kalpakkam in Tamil Nadu. I had often discussed about this bioshield demonstration project with Dr. Kalam earlier. So, during one of my visits to Delhi, when I appraised him about VVA, he showed keen interest and gave many useful suggestions including preparing an action plan. This helped us a lot and we started preparing a project report for such a holistic plan.

When Dr. Kalam was again visiting Vadodara as the Guest of Vivekananda Centre at Rama Krishna Mission in 2012, I requested him and he agreed to come and address a large gathering of Volunteers and students and release the project report.

He interacted with hundreds of school children of Vadodara at the function. This was indeed an exciting time. What was amazing was 2 days before his visit he had sent me the full text of his talk titled "River Revival - VVA", in advance. His speech as usual, was so inspiring and at the end he asked everyone to join in oath taking.

"Environment Oath"

1. I will plant and nurture ten trees and will ensure my parents, my sisters, brothers and my neighbours also plant ten trees each. I will be an ambassador for tree mission in my locality.
2. I will keep my house and its surroundings clean and use products which are bio-degradable to the extent possible.
3. I will promote a culture of environment friendliness, through recycling and conservation of water and other recyclable materials both at home and school.
4. When I take a professional career, I will take a decision with respect of organizational processes which protects the environment and preserves the bio-diversity.
5. I will encourage the use of renewable energy to the maximum extent possible.
6. I realize, with every tree I plant, I am contributing in good weather making.
7. I will spread the awareness about the need to preserve the environment in my home, in my locality and among my student friends".

He concluded with - "My best wishes to the Vaho Vishwamitri Abhiyan which is a voluntary, grass-root movement for reviving Vishwamitri river and its watershed from Pavagadh to Gulf of Khambhat including villages, towns and Vadodara city. It is indeed a big mission. It needs big thinking. Of course it needs big action. I am sure, your mission Vishwamitri Abhiyan will succeed with bioshield and eco-friendly sustainable system".

Before departing, Dr. Kalam said, 'Vishwamitri is Possible' and declared that VVA should be a Model Project for the whole country. Our entire team's enthusiasm increased several folds due to this visit and we went on to develop Seed Ball Technology (coating different types of seeds with minerals, bio fertilizers etc 'seed balls' which can be thrown from distance in hills, ravines etc.) for Plantation. This is a fast way for developing plantation and bio shields. As a matter of fact, he was again to visit Vadodara, to address the International all ReligionCongress organized by Ramakrishna Mission in August 2015 and we had arranged two Seed Ball Camps on the bank of River Vishwamitri during that time. The shocking news of his passing away in another corner of the country just a few days before this event brought tears to our team and students, Volunteers in Vadodara.

Avul Pakir Jainulabdeen Abdul Kalam grew up from a modest background to be a great and inspiring leader of the country. His life story is a journey of achievements and all round development of a human being touching millions of hearts. Like Einstein, he was a pantheist - a believer in all Gods and one who saw God' in all natural things – and worked for the good of all.

Dr. Kalam's vision and life will be one of the greatest assets for future generations. I quote from one of his less known books – "The Tree of Life:-.

"As a young citizen of India,

Armed with technology, knowledge and love for my nation,

I realize, small aim is a crime.

I will work and sweat for a great vision,

The vision of transforming India into a developed nation,

Powered by economic strength and a strong value system".

Where your talents and the needs of the world cross, there lies your purpose.

— Aristotle

21

Remembering Dr. Kurien

"Where is your Dhoti, Dr. Mehta ?" Dr. Varghese Kurien surprised me as I entered his office at IRMA (Institute of Rural Management Anand). That was April 2000, soon after I had taken over as the Vice Chancellor of the erstwhile Gujarat Agricultural University (GAU) and had decided to call on him during my visit to the Anand Campus. (At that time GAU was the largest University in the Asia with four main campuses – Dantiwada, Anand, Junagadh and Navsari and about 65 experimental farms / stations spreaded all over).

Earlier, I did have some interactions with Dr. Kurien. When at GSFC Ltd., we had introduced Nylon – 6 Engineering plastic. I had discussed with my team of Research and Application Development Engineers about multi layer packaging of dairy products where Nylon – 6 can have a good market potential which we used to discuss with Amul. Also, our work on Bio Fertilizers and Bio Pesticides had indirect interest to them. So, when suddenly I was offered the position of Vice Chancellor of GAU and was in a dilemma, I went for his advice. After all, he had held the same position earlier (in addition to AMUL). I explained about the offer and said that I was not too sure about accepting it. He, on the contrary, encouraged me to

April 2000 : Old guard and New guard
Vice Chancellors - Gujarat Agricultural University

accept it saying "you can do a lot !" I just murmured," But I am not an agriculture graduate. I am a Chemical Engineer!" To which he jokingly said – "So ?. I am a Metallurgical Engineer and now I am a Dudhwala!"

Eventually, I accepted and so when I visited Anand soon after, the first thing I did was to call on him. And as I entered his room, he asked "so, where is your Dhoti?". I said – "What do you mean, Sir? I never wear a Dhoti!" He said, "As Vice Chancellor of Agri University, you should wear a Dhoti!". It was the beginning of my many meetings and listening to his funny remarks and stories.

EARLY YEARS

Varghese Kurien was born on 26th November 1921 in Calicut, Kerala and went on to do B.Sc – Physics at Loyola College –Madras and then Engineering at Guindy Engineering College. He was an all rounder- Tennis, Cricket, Boxing ! Soon after, he was selected by TISCO Jamshedpur as graduate apprentice but left for USA when he got a scholarship through a Government Scheme. He wanted to do metallurgy or Nuclear physics – but was told that he was selected to do Dairy Engineering – at Michigan State University. When he returned to India and reported to Ministry of Education, they asked him to go to Anand. 'Where is this Anand?' he asked. Very unwillingly, he came to Anand. Under the loving care of Tribhuvanbhai Patel – a simple honest Gandhian who commanded a great deal of confidence of Sardar Patel and Morarji Desai – Kurien joined Dairy Cooperative Society that laid foundation for the White Revolution!

Remembering the old days in the sleepy town Anand where he could not rent a house – A non-vegetarian Christian and a Bachelor - he started 'home' in a modest garage !

The first 'Home' - The Garage House with his mentor Tribhuvanbhai Patel and wife Molly

STORY TELLING AND SHARING.

Dr. Kurien was an interesting story teller and had a great sense of humour. During my frequent visits to Anand Campus, I would meet him in his office ' for ten minutes' which would invariably last much longer.The topics would vary from his early days in a Garage house at Anand to confrontations with Politicians and Babus and his own days at University (GAU) and new developments. His stories would usually end with suggestions for being bold and forth right. For me these were very important lessons.

During one such session, I mentioned how I had rebuffed a senior minister in the State Government, who wanted to interfere with the University administration by asking for transfers of some officers. I had conveyed that "it will be strictly as per research project needs of a committee I had formed". The minister, however, telephoned me again and said, "there are so many MLAs right now in my room – asking for such transfers. What will I tell them?"I retorted, "tell them that this VC is mad and does not listen to me!" He was shocked "what ?" – "yes" – "I said you heard it right". Tell them that this VC is Mad!" and hung up. When I narrated this incidence to Dr. Kurien, he was much amused and patting me on my back said, " in this position, always keep Resignation Letter in your pocket " (some how I did not lose the job !). He however suggested a more safe and practical way he has learned with time was to say "Joisu – Joisu (will see – will see !). He further went on to narrate his several encounters in Delhi and elsewhere.

One such story he told and it is now well known was his encounter with late Jagjivan Ram – then the powerful Minister of Agriculture – GOI. "He called me once in his office and said "Dr. Kurien, I want you to build a private dairy in my constituency" – Kurien." I will not do a private dairy. I am here to build only cooperative dairy". – The Minister was very upset and later even sent a note to PM office to 'remove Kurien'. By that time, sensing trouble Dr. Kurien had sent a word to PM office about the incident and it seems Smt. Indira Gandhi returned the file with a remark "Do not touch Kurien, Leave him alone". (This incidence is discussed in detail in his book "I too had a dream" – Roli Books 2005). Narrating this he said, "instead of rebuffing, I could have just said "will see – joisu!"

He would also caution "not all your agriculture scientists are 'SIDHA' – straight forward! Even a man like Swaminathan had lots of trouble in ICAR! So, be careful!"

These interactions at Anand with him were extremely useful lessons for me all throughout my GAU days!

VIDYA DAIRY ANAND

Anand campus of GAU had always given great support to AMUL, NDDB and even IRMA. One glorious return gift from AMUL to GAU was jointly setting up Vidya Dairy, a students' dairy. This 1 lac litre capacity model dairy is operated mainly by 3rd and 4th year students of Dairy Technology College under 'Earn While you Learn' program. It has been a huge success as not only these trained students have been ingreat demand when they graduate but are very happy to earn some money while studying! Besides, the dairy ran as a profit making venture and several new varieties of cheese, icecream etc were developed and scaled up for AMUL!

Vidya Dairy – Students' Dairy – GAU - Anand

Dr. Kurien would very proudly mention this. With this experience of Vidya Dairy, I sat with my GAU team to work out similar plants in all its campuses. We decided to set up 4 plants on Earn While you Learn Model : (i) Spice Processing at Sardar Krushi Nagar – Dantiwada (ii) Mini Oil Mill near Junagadh (iii) Fruit Processing Unit at Navsari – Campus and (iv) Mini Pulse Mill – Anand.

I often shared the progress with him and he would endorse "the right thing to do – should be taken up at all Universities….."

OPERATION FLOOD

The purpose of Operation Flood was to replicate Anand in a number of Indias major milksheds. Unlike many, Dr. Kurien understood the importance of Marketing and would often remark "we talk about increased production of milk but not market! Marketing is more important !" In designing Operation Flood, the development of women and men farmers was the central theme. The production of milk is kept in the hands of small farmers. The entire structure fits with the spirit of India, of a decentralized rural economy. The cooperative never encourages huge, bureaucratic system. Today, 11 million farmers in 22 states across the country own 200 dairy plants handling 18 million litres of milk a day.

Operation flood made our country self-sufficient in milk and the largest producer in the world. By 1975, all imports of milk and milk products stopped. In 1955, our butter imports were 500 tones / year. Today, our cooperatives alone produce 12000 tons / year.

IRMA

During one of our meetings, Dr. Kurien mentioned about his interactions with his old friends – Vikram Sarabhai and Ravi Mathai (Director IIM-A) and the genesis of planning and setting up of IRMA (Institute of Rural Management – Anand).

IRMA was created from the budget provision in NDDB of Rs.50 Lacs for training of human resources to manage cooperatives. The idea ultimately evolved into IRMA and a budget of about Rs.10 Crores was set aside. IRMA has a beautiful campus – adjoining GAU Anand campus and was established with a great vision and commitment.

Today, many of IRMA students are in senior positions with cooperatives or NGOs and some in large companies. As the Hon. Chairman, he planned and managed it with the same commitment and dexterity which he showed in building AMUL and NDDB. Often there is a tendency even by senior bureaucratsand motivated people to invade and play power politics when a good institution is created. This hurts the institution builder. He resisted that strongly. He would say, "People may call me cold blooded, but I am very firm in setting correct procedures"

LAST DAYS

Eventually, he left IRMA and continued to live in Anand which he called his 'Home'. Like many, I tried to keep in touch with him and as his health was failing, it was his wife Molly or Mollyben – as she was affectionately called – who would answer the phone calls. Later as I was visiting Anand to attend a felicitation function, I called her that me and my wife would like to personally come and take his blessings. She said "you are welcome, but he is not at all well and he may not even recognize you". We went anyway and found him waiting with the same old smile and warmth, though he looked thin and pale. He said "Congratulations", pressed my hands and posed for a group photo willingly. Mollyben was surprised and happy. That was my last meeting with him. He passed away after a few months.

At Home – the last meeting

I have often thought about the working styles of scientists, technocrats and professional managers in different organizations and at Government levels. It seemed to me that there are three types of people – Those who cut themselves to small size and submit to the system. The second group are those who find a way silently around the obstacles in the existing system (examples Dr. Vikram Sarabhai, Prof. Swaminathan !). And the third type are those who fight their way through and achieve big things based on strong principles, and a single mission. Dr. Kurien belonged to this class.

Nation celebrates his birth day (26th November) as National Milk Day. His life and achievements are one of the most glorious chapters of the post independent India. As Ratan Tata said "one cannot help wonder what India would be today if we had a thousand Dr. Kurien with his type of vision and with similar commitment, dedication and national spirit".

"Your success did not create your optimism, your optimism led to your success"

— *www.gatesnotes.com/2017*

22

Sustainable Agriculture – A Base For Rainbow Revolution

In contrast to the traditional universities, agricultural universities have been entrusted with the Herculean task of technology dissemination at the grass root level such that the package of practices evolved by the agricultural scientists should reach out the end-users, viz. farmers , for adopting in the field, with consequent increment in the farm productivity and farmer's income.

This unique mandate has been assigned to agricultural universities on the pattern of land grant colleges of the United States. Such vision and trust, envisaged by Pandit Jawaharlal Nehru and his devout lieutenants, herald creation of state agricultural universities (SAUs) in the early 1960s. The key role played by SAUs in the first green revolution and agricultural human resources development have been regarded as the milestones in the history of India post-independence, that made the country self-sufficient in food production and, in turn, to self reliance in milk, vegetables, fruits and inland fish production.

Today, India has set fourth into commanding position as a global economic power, with primary and secondary sectors, viz. agriculture and industry, contributing modestly. At the current pace of the growth rate, I would not hesitate to predict that by the year 2020, we will accomplish the top lines both in terms of productivity and sustainability, by transforming hitherto deemed liability of population into a formidable asset of human resource capital.

HUNGER-FREE WORLD

Agriculture is a vital development tool for achieving the Millennium Development Goals that call for halving by 2015 the number of people, suffering from extreme poverty and hunger. We also note that more than 50% of the people in the world are small and resource-poor farmers and another 20% rural land less. The challenge is to transform this concentration of poverty in agriculture into an opportunity. The Food Price Index of FAO, based on export prices for international trade food items, climbed 37% in 2007.This, was on top of 14% increase in 2006! Thus, a startling change is unfolding in the world's food markets. Soaring fuel prices have altered the equation for growing food and transporting across the globe.

The Indian economy has been growing at about 8% per year since 2002 and even today the growth rate of more than 7% appears impressive. At the same time, nearly 48% of the children are under-nourished and the challenges in the coming years are bigger. We must keep this paradox in mind in our future planning.

RAINBOW REVOLUTION

What kind of dream you have for India? What kind of world, in which India will be playing a very important role, do you dream of?

One of our greatest achievements of India in last few decades has been the Green Revolution. From a starving and hopeless country, dependent on imported food grains, we have become a food surplus agriculturally advanced county. We have also ushered in White Revolution – by becoming the largest producer of milk in the world.

New developments in biotechnology, IT, space sciences, energy, new materials, health sector and management will open extraordinary opportunities in the coming decades. We should be prepared to bring several new revolutions, which could give us all-round improvement and a quantum jump. Such a rainbow revolution will have:

- Eco-friendly and sustainable agriculture;
- Blue revolution: seawater farming and fisheries;
- Brown revolution: making deserts and semi-desert bloom;
- Black revolution: biofuels;
- Multicolored revolution: horticulture and food processing;
- White revolution; and
- Drop of water revolution: Water use efficiency for agriculture, i.e. more crop per drop.

All these revolutions can help to remove hunger and poverty, improve health, have cleaner and sustainable environment, minimize inequality and create new opportunities for a better and healthy world. It can transform India into a developed country and have leadership position in the world.

Prior to the Green Revolution, farmers in India were largely dependent on monsoon, organic matter from farm animals and previous year's harvest of seeds. There were failures and successes. Beginning 1967, the Green revolution changed all that. High Yield Varieties (HYV), chemical fertilizers and pesticides and irrigation helped to raise agricultural output and within a short span of 10 years, India become food self-sufficient.

The Green Revolution changed India's position from a food-deficient to self-sufficient state. But it also exerted considerable pressures on the agri-ecosystems, owing to extensive use of agrochemicals. Also the small farmers, which are a large percentage, were mainly left the side.

ECO-FRIENDLY AND SUSTAINABLE AGRICULTURE

According to a study carried out by the World Watch Centre, around 70,000 chemicals are in everyday use all over the world. Awareness about their effects has, however, lagged far behind. An IMCR (Indian Council of Medical Research) study revealed (Garg, 2006) that 51% of the food items were contaminated with pesticides. Centre for Science and Environment, New Delhi found contamination in bottled water with agrochemicals like DDT, lindane etc.

Studies also indicate that only a small percentage of agrochemicals used – whether as fertilizer or pesticide – is effectively utilized, with the rest going as waste in soil, water and air. Indian farmers spend Rs. 2 lakh crore on agro inputs including agrochemicals, seed etc. Further, in India, nearly Rs. 50,000croregoes as chemical fertilizer subsidy. Fertilizers use efficiency is drastically declining, e.g. in mid-60s, 1 Kg of chemical fertilizer gave 19-Kg of food grain, which has come down to less than 12 Kg now. A study by IFPRI (International Policy Research Institute, Washington) (Gulati, 2007) has shown that $ 1 spent on subsidy gives $ 0.53 return only, as against $ 6.9 for R & D and $ 3.2 for a rural road. There is, therefore, an urgent need to find an alternative system which is more productive, efficient and eco-friendly.

An ecological production management system that promotes and enhances biodiversity, biological cycles and soil biology activity is the need of time. It has to be based on minimum use of agro-inputs. Alternativebio-agro inputs such as Bio-pesticides, Bio-composts, Bio-fertilizers and Bio-control agents will emerge as alternative an significant inputs. Scientific understanding of multi-microbial technology has considerably increase in recent times. Consortia of micro organisms (Mehta, 2007) consisting of various combinations of Nitrogen fixing, Phosphate solubilizing and with bio fungicide activities show excellent field results. Strains of micro organisms, which can hasten the process of composting of organic residues, have been applied. Such microbes are cellulolytic and lignolytic type. Homogenised fungal culture *Trichurus spiralis*, *Paecilomyces fusisporous*, *Trichoderma viride etc.* give accelerated bio-composts of uniform quality.

THE 20-20 MODEL

It is now worth noticing that a 20-20 Model, which helps to reduce input costs by 20% and at the same time improves farm productivity by at least 20%, has been demonstrated in different locations and for different crops. Also, we must emphasize an excellent model of NGO-University-industry collaboration, in which NAU was the key member, helped in evolving this model through field demonstrations. During the recent debate on BtBrinjal, such a model was discussed as easily implementable and farmer friendly one, which should be our high priority. It is time to take such a model for the benefit of a large population of farmers and perhaps a state like Gujarat can take leadership.

VALUE – ADDED AGRICULTURE

This is the need of the time and is the Key factor for lifting up the level of the society. Several examples like value - addition chain points to the fact that the farm produce could be raised to much higher value through the input of good technology and process integration. Some examples like Castor oil to Bio – Nylon (from Rs.50/Kg. to Rs.1100/Kg.) Spices and foods processing (value addition by 10 to 50 times), milk products, show the case.

CENTRE FOR APPLIED MARINE BIO-TECHNOLOGY

The problem of Global Warming will have one of the greatest impacts in the coastal areas. Closely associated with this problem, is the concern for water storage and salinity. Coastal areas also have poorest of poor population. On the other hand, development like integrated farming involving aqua agro complexes, development and cultivation of salt tolerant plant (halophytes) as well establishing coastal bio shields can greatly help in mitigating natural disasters, environmental improvement and economic returns. A number of groups, including the scientists of NAU, have done commendable work in this areas.

Einstein said that 'today's problems can be solved by tomorrow's technologies only'.

Looking at the need of the future, there is a case to set up a Centre of Marine Bio Technology Research and Application close to Navsari. Such a Centre can be developed to provide leadership in areas like :

- Saline water agriculture, including salt tolerant varieties ;
- Integrated Aqua – Agro complexes;

- Bio-Shield with Biodiversity plantation for disaster mitigation and improvement of eco-environment;
- Algae based bio fuel products for health care and plant growth hormones, *etc.*;
- Coastal eco cities, *etc*;

I am sure that such a center can be established under Public-Private Partnership mode and can be a fitting tribute to the great people associated with Navsari and Dandi like Mahatma Gandhi, Dadabhai Navrojji and Jamshedji Tata.

NEW CHALLENGES FOR AGRICULTURAL EDUCATION

At the time of independence, India had 17 collages for agricultural and allied sciences. Soon after Independence, our urgency of bringing about rapid increase in food grain production necessitated reexamination of the existing pattern of agricultural education. Thus, the Government of India appointed a University Education Commission under the Chairmanship of Dr. S. Radhakrishnan to review higher agricultural education in order to suggest ways for fulfilling the future requirements of the country. The commission recommended that agricultural education be recognized as a major national priority, so that the country is able to meet her food requirements. The first Indo-American team in 1955 made a study of agricultural research and education in the country and in its report recommended strengthening and reorientation of agricultural education, including veterinary education. The committee further recommended that the development of rural universities should be assisted by substantial grant in aid from the Centre, which will ensure autonomy and efficiency in operation. The first agricultural university thus came into being during 1960 in Pantnagar, UP. In 1964, an Agricultural University Committee headed by Dr. Ralph W. Cummings, recommended the establishment of Universities in 6 States with respect to work in agriculture, including animal science and allied sciences basically on the pattern of Land Grant College System of USA.

The ICAR Committee under the Chairmanship of Dr. M.S. Swaminathan constituted to examine the Agricultural Education System in India, has observed "Agricultural Education has to get out of its mould of a formal organized rigid framework and has to take the role of continuing education where the education process is adjusted to the needs of illiterate, unskilled farmers and farm house holds. This would imply that the individual farmer should have access to the agricultural facilities at different stages of his life cycle and within a period of the year in which availability of time to him is higher, it would also mean that educational process has to be made more directly related to the local conditions and to be made more meaningful, shall have to be a part of work and employment culture of the area". We have to keep this aspect in mind in planning our system and implementing our agricultural programs.

EPILOGUE

Dr. S. Radhakrishnan has said that "Science has achieved more for the emancipation of the masses than the wisdom of the sages". We are at a very interesting phase of human history, where scientific knowledge and its application can be the vital tool to help solve our major future problems. Thus, we should build the future programs on the strong foundation already laid through our dedicated efforts and focused research and academic programs based on partnerships with institutions, private sectors, NGOs and the Key stake holders.

(Based on the Convocation Address delivered by Dr. M.H. Mehta on Feb. 15, 2010 at Navsari Agricultural University.)

Actions today – not promises for tomorrow.

— *M.S. Swaminathan*

23

Our Soils – Our Waters and COP22

OUR SOILS – OUR WATERS & COP22

In the last several decades many international forums and institutions have discussed and agreed that food and water security are the most important issues for the future. The damage to the ecology directly affects the food, land, water, health and the livelihood. Agricultural progress serves as the best way to fight poverty, hunger and livelihood problems.

Though in the last few decades, billions of pages about hunger, food and water security have been written, thousands of conferences have been organized, the urgency of action plans and creating holistic models is the most important need. For this, we must require bold reforms and action oriented programs which can reach the smallest farmers and give an optimistic signal rather than a despondent picture. Developmental models that are holistic and close to nature, sustainable and based on inclusiveness considering 'unto the last' women and man are the need of the day.

Land and water together with biodiversity and atmosphere conditions are the foundations for sustainable food security. The ecological damage to these foundations are leading to lowering of productivity and livelihood. This will further aggregate with the threat of climate changes such as temperature, rain falls and sea level rise. As per the exercise by World Food Program, it is indicated that the green revolution states like Punjab and Haryana may become food insecure in another 20 years. Food insecurity is measure in terms of land degradation and salinization, ground water tables and green cover loss.

---

* Based on Key Note address at Global Water Meet for Climate Change Adaptation – Agrarian Perspectives : 24 – 26 OCT. 2016 – UAS – DHARWAD – INDIA

OUR SOILS

The year 2015 was declared ' The International Year of Soil'. This is understandable. A healthy soil is key to farm productivity. In India, out of 306m/ ha reported under various land uses, about 141 m. ha is not sown area and a large portion of this has undergone degradation. Besides a huge area is affected by water logging, ravines, shifting cultivations etc. In Table 1, the state of key health indicators are given. This is a typical example extendable to most countries in Asia and Africa. In fact, African farmland is losing fertility at an alarming rate.

Table 1: State of Key Soil Health Indicators in India

Sr. No.	Parameters	Status
1.	Soil Organic content	Low SOC, ranging 0.2 to 0.5% in most soils.
2.	Alkali Soils	~ 3.8 m ha – restricted production due to salinity.
3.	Acidic Soils	~ 6.5 m h – toxicity and nutrient deficiency.

Soil Organic Carbon (SOC) content is the most important determinant of good soil health and support systematic microbial activities. Microbes are responsible for providing food to the world and we need to feed the soil microbes in the form of organic carbon. Putting food for microbes will provide food on our table. The soil microbe biomass i.e. bacteria, acitomycetes, fungi, humus etc. are the chemical engines that perform the organic matter degradation. The sources of humus and SOM serve as reservoirs of living organisms, which participate in the mobilization of plant nutrients and overall help give a healthy soil. Table 2 gives a brad list of overall soil fertility criteria. This is a general guidance which with the Soil Health Card initiative can significantly help in taking actions for a healthy soil.

Table 2: Soil Fertility Classification (Ref. 3)

Nutrient %	Amount of nutrients %		
	Low	Medium	High
Organic Carbon	< 0.5	0.5 to 0.75	> 0.75
Potash (Kg/Ha)	< 140	140 to 280	>280
Sulfur (ppm)	< 10	10 to 20	> 20
Zn (ppm)	<0.5	0.5 to 1.0	>1
Fe (ppm)	<5	5 to 10	>10
Cu (ppm)	<0.2	0.2 to 0.4	>0.4
Mn (ppm)	<5	5 to 10	>10
Boron (ppm)	<0.1	0.1 to 0.5	>0.5
Mo	<0.05	0.05 to 0.1	>0.1
pH	<6.5	6.5 to 7.6	>7.6
Dissolved salts	<1.0	1 to 2	>2

REGENERATING THE SOILS

Soil formation is the result of synergy between clays from mother rock and humus. From the organic debris soil degradation and erosion in the reverse of soil formation, i.e. instead of fusion of clay and humus there is a separation and degradation of these elements. Less fauna means more erosion / degradation. If you destroy fauna by excess chemicals you have no oxygen4 and therefore anaerobic bacteria increase which converts nitrates back to ammonia – so you need more fertilizer.

Soil microflora consist of 4 groups – Algae, Fungi, Filamentous bacteria or acetomytes and Bacteria. Bacteria transform nitrogen to Ammonium and then to nitrate. Bacteria also act in two ways - (i) Oxidation (oxides of P, S etc and turn them to phosphates, Sulphatesetc which can be assimilated by plants.(ii) Chelatation i.e. attach a substance to complex organic molecule so that it can be assimilated as micronutrients (about 20). Membrane : Molecule like Nitrate has high salt concentration and has osmotic force. Due to this, the plant cell has to bring in more water inside the cell (This is called 'Nurgar Pressure') so the plant becomes full of water. Also there is 'disequilibrium' as Ca, Mg etc with two positive charges cannot permeate membrane and your food is not balanced. In temperate regions, soils have high reaction exchange capacity but in tropical regions it is very low. But if you have soil with humus or fauna you have microbes to produce nitrates and other nutrients and fauna to produce galleries for free flow of oxygen.

COMPOSTING FOR SOIL HEALTH

Composting helps (i) create 1 hot fermentation which pasteurizes organic matter. (ii) Much higher quantity of humus is produced (eg 30 tons of manure gives 3 tons of hums but when composted it gives 6 tons humus).

Recycling of organic waste to maintain soil fertility and humus content is of paramount importance. As discussed elsewhere, in India more than 500 m tons of crop residues are generated very year and nearly 90 mt / year is burnt on farms. Soil is the stomach of plants teeming with millions of hungry microbes.We need to feed them carbon to enable to feed us.

The composting process is :

- Biological (Run by microbes)
- Works on organic feed stocks
- Is extension of natural decay – controlled – managed and accelerated

- Predominantly aerobic
- Depends greatly on C:N ratio and moisture.

A field study in New South Wales – Australia found that the use of compost saved 14,000 gallons / acre of water. Similarly, Rodale Institutes farming trials concluded that with composting (and no chemical fertilizers) even for 10,000 medium scale farmers in USA will store so much carbon in soil that is equal to taking almost 1 million cars of the road !

SOIL HEALTH CARDS SCHEME AND SOIL FERTILITY AMENDMENT

While preparing Agro Vision for Gujarat in early nineties, a great deal of emphasis was placed on soil health cards and soil improvement needs. As the Vice Chancellor of Gujarat Agricultural University, formed a team to initiate the mammoth task of soil analysis for different regions. Recently, in 2014 Government of India announced a very ambitious plan for Soil Health Card for the whole nation. Initially it has aimed to cover about 140 m farmers. It has also been mentioned that the ideal N:P:K ratio is 4:2:1 and the correct ratio can save farmers upto Rs.50,000 / year for each 3 acre plot. Unfortunately, for Punjab and Haryana soils this ratio is 20:6:1 (or 3:1: 1/6). Also the fertilizer subsidy in urea takes nearly Rs.48,000 Crore and that the extra use of urea is not helping at all and is a big financial loss.

For many, however, a Soil Health Card means a major focus for optimum application of chemical fertilizers and micronutrients. This perception needs to be broadened. We have to look at the overall soil health and its improvement practices. As pointed out by Uphoff "Fertility does not depend upon how much you put into the soil, so much as what the biological conditions in the soil itself are – the fungus, the bacteria , the protozoa doing their functions in a very active system".

So when a farmer floods the soil, it becomes hypoxic and does not get oxygen. Most bacteria, and all fungus need oxygen. But when you add organic matter to the soil, roots and organisms benefit. Plants grow in symbiotic - the real agents for growth are soil and microorganisms. There is not enough appreciation of what natural system can do and how we can improve upon.

In the SRI (System Rice Intensification) – the emphasis has been on improving matter and ensuring aerobic soil environment through controlled application of water and using younger seedlings. Nurturing their roots and wider spacing demonstrated a huge jump in rice productivities (20 to 50%) and water saving (30 to 50%). It is estimated that in India alone the total farming area under SRI/SCI is already about 2 m ha !

COP21, COP 22 AND 4 PER 1000 INITIATIVE

Conference of the Parties (COP) to the UNFCC (United Nations Frame Work on Climate Change) has been making significant efforts to mitigate the adverse climate change efforts. At COP 21 in Paris, a landmark agreement was reached to combat climate change and accelerate actions for sustainable low carbon future. COP 22 at Morocco will focus on short, medium and long term adaptation and amelioration strategies. Water and agriculture will form the extremely important foundation for this program. It will be crucial for the farmers to adopt new technologies and adapt to the new and adverse conditions.

4 per 1000 Initiative

IFOAM – Organics International has joined the "4 per 1000 Initiative" which aims to improve the organic matter contents and promote carbon sequestration in soils through application of agri practices locally both environmentally, socially and economically.

"Agro ecology is at the heart of 4 per 1000 initiative" - noted Gabor Figeozy of IFOAM and French Minister of Agriculture Stephane Le Fale. 4/10000 (= 0.4%) is the annual growth rate of soil carbon stock. 4 per 1000 aims to adapt agri practices with the goal of storing carbon legume efficiently in the soil. An annual increase of 0.4% organic matter in soil is enough to compensate for global emission of green house gases.

The main steps are :

- Reduce chemical fertilizers
- Use legume in crop rotation
- No till (or minimum till) practice
- Agro forestry
- Plating more cover crops and not leaving the soils.
- Changing diets of cattle

4 per 1000 for Food Security & (Sequestration in soils)

The amount of carbon increase in the air is nearly 4.3 billion tons per year. It is estimated that World's soils contain 1500 billion tons of carbon in the form of organic material. So, if we increase carbon in the soil by 0.4%, we can halve the annual increase of CO2 in air.

Therefore, with 4/1000, we will have more fertile soils and reduced carbon in the air.

I recall an interesting anecdote – while talking to farmers group in the semi arid state Rajasthan (Western India), I was talking to them about Green House gases – the water and agricultural crisis etc. An old farmer raised his hand and said – "Sir, the problem is simply this – you want carbon in the soil but it has escaped in the air!"

The eco agricultural practices to improve soil health has close link with the management of farms agro waste and converting it to high value bio composts. This also has a great impact in improving water use efficiency. This, together with crop rotation, integrated farming and the use of multi microbial bio fertilizers will offer immense scope of meeting the millennium goal of hunger free and sustainable world. As Lavosier has said, "Agriculture is sustainable only if it restores to the soil what it takes from it".

OUR WATERS

The issues of water are extremely important for food security, human health and the environment.

There is a close relationship between water gap and food gap. A simple fact is that 97.5% of total volume of water available in this planet is saline. Of the remaining, a great deal is locked up in ice caps and not easily available for human use. Of the remaining which is accessible, 80% come in bursts as monsoons! Thus, the quality you are left with for use is small.

As the demand for water outstrips supply, river basins will get water stressed and scarce and ground water aquifers will reach critical conditions, creating larger socio – economic imbalances and food insecurity.

The Water Vision Document 2025 stated that "Poverty will be eradicated and living conditions of all people will be uplifted to sustainable levels of health and well being through integrated development and management of water resources." It further listed out Vision elements – some of the main are :

WATER FOR FOOD – VISION ELEMENTS

Water vision cannot be formulated in isolation. It has to be closely related to the general economic vision of the region.

Develop technology package to increase water use efficiency in irrigated agriculture and land productivity in rainfed agriculture.

Use improved information technology to disseminate information to farmers, both men and women, rather than depend on inefficient extension methods.

Need clear demarcation and balance between roles of public and private sectors; water cannot be left entirely to market forces.

Provision of accountable, efficient and effective water and sanitation services to all consumers enabling access to safe, sustainable and affordable services for better quality of life by 2025.

Zero morbidity and mortality due to diseases associated with unsafe and inadequate water supply and sanitation services.

River basins be taken as environment planning and management units.

Watersheds to be conserved.

Flood protection must be dealt with at a regional level in the case of transboundary river systems.

Regional cooperation in flood and drought management.

There should be a regional water policy.

Attention must be paid also to quality of water, not only quantity. Establishing and strengthening pollution control institutions.

Institutions for protection of aquatic habitats and ecosystems.

WATER BALANCE CHALLENGE

India gets nearly 4000 billion m^3 (BCM) of water from all natural sources. Out of this as seen in Fig. 1 a large quantity - 1500 BCM flows into the sea. The available water is only 1100 BCM. Over 70% of the available water is used for agriculture. Every cubic meter of water produces 300 gallons of crops. This is a case of poor water use efficiency (WUE) which compared to developed countries is very low. There is ample scope for more scientific cultivation using less water and loss less land such as drip irrigation.

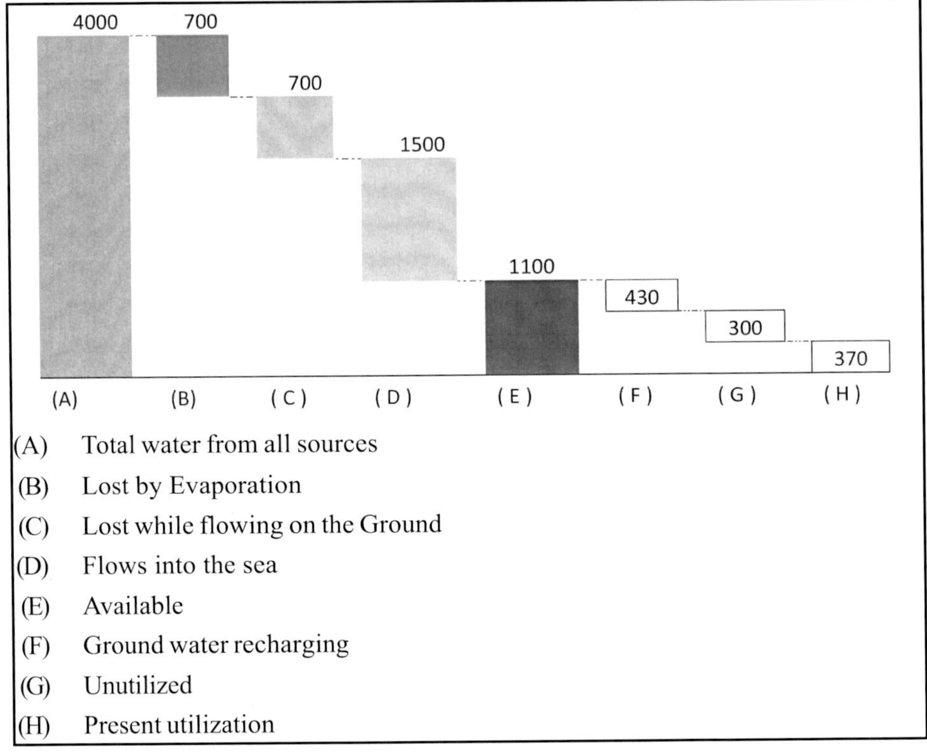

Fig. 1: India's Water Balance details in BCM (Billion m^3)

GUJARAT GREEN REVOLUTION CO. (GGRC)

Though for long, the importance of improving water use efficiency has been appreciated – execution and implementation program is a major challenge. In late nineties, we established a company through the joint contributions of GSFC Ltd. (Gujarat State Fertilizers Co. Ltd.), GNFC (Gujarat Narmada Valley Fertilizers Co.) and GAIC (Gujarat Agro Ind. Co.) to promote work on efficient and value added agriculture and water management. Earlier the company was named as Gujarat Agri Processing Co. Ltd. (with me as the founder MD of the company) which was later renamed as Gujarat Green Revolution Co. As noted by Dr. Kalam in his book, "GGRC was established to help farmers with investments and give them access to technology (for WUE). The farmer had to contribute 5% while 45% come as loan from NABARD and balance 50% as subsidy from GGRC. It was reported that at the end of six years the farmers' agricultural income doubled and is growing further. It also help contribute to the overall agricultural growth of Gujarat to 9% as against the national average of

2%. Thus, proper management of natural resources is vital not only for increasing agricultural yield and income but also for mitigating adverse impacts of water scarcity".

ECO AGRICULTURE AND WATER SCARCITY

Developments in Eco agricultural practices can play a very valuable role in improving water scarcity scenario. It is known that most irrigation schemes are highly inefficient and eco agriculture / organic farming need much less water and have a better WUE. It is reported that use of compost in New South Wales – Austrian fields, there was a significant saving (upto 1,00,000 gallon per year) when compost and organic practice was taken up.

Similarly, as shown in Table, the use of bio fertilizers alongwith drip irrigation gave a substantial saving of water and improvement in Banana productivity.

Table 3: Eco – Agriculture & Water Use Efficiency
Trials of Fertigation on Banana Crop
At GAU – Navsari& Farmers' Field, Dakor (Gujarat)
(Ref. Report – More crop per drop
S. Raman, RPS Ahlawat and M.H. Mehta
Gujarat Agri. Uni. (GAU) Report 2002)

Parameter	Flood Irrigation+ Conventional Fertilizers	Drip Irrigation+ Conventional Fertilizers	Drip Irrigation+ Liquid* Biofertilizers+ Conventional Fertilizers (reduced)
Water Consumed ha-cm	250	150	140
Yield ton/ha	50	70	94
Returns Rs/ha	95,000	1,35,200	184,247

We also see that between the two extremes of all chemical farming and organic farming a 20:20 model of Eco Agriculture is proving very practical and effective. This applies new generation agri bio inputs (Bio Fertilizers, Bio Pesticides, Bio Composts so that farm productivity increases with lower input costs in a sustainable manner.

SALINE WATER AGRICULTURE

Salinity is a significant stress for plant and has been shown to account for upto 70% loss in the crop yield. There has been considerable interest in saline water agriculture by use of advanced soil management and irrigation technology and biological approach to improved species of plant, introduction of desirable agronomical practices and development of halophytes. Saline Water Agriculture will form an important future strategy and has been discussed in a separate chapter enclosed.

Globally, our water scarcity problem is closely associated with the problems of salinity, drought and quality of water. These water stresses are important to understand and take up an exploration of new technology to solve them.

PRIORITIZING WATER AT COP22

On July 2016 an International Conference on Water and Climate was organized as a preparation for COP 22. The conference underlined that water sector represents potential solutions to climate change in addition to being particularly climate – vulnerable sector. Particularly the ministers discussed African mobilization around the effects of climate change, the sanitation sector and cooperation with the private sector.

As we know there are increasing problems of soil salinity and desertification, water logging, depleting ground water tables, water pollution and loss of forest and dio diversity. An integrated approach for : Water conservation, creating bio shields particularly around water bodies like rivers and ponds, non-polluting agri bio inputs like Bio fertilizers, bio pesticides, conversion of farm agro waste to bio composts and eco-agricultural practices et are the actions that need to be demonstrated and multiplied.

PUTTING IDEAS INTO ACTIONS

We have so far discussed about our 'Soils' and our 'Water' and the key role of related technologies with respect to COP 21.

In his book from Rio to Johannesburg, MS Swaminathan talks about "Action Today and not just Promises for Tomorrow". It has also been felt that COP 21 at Paris left many feeling (Ref. Sunita Narayan) that the umbrella group led by some rich polluters, who have always been in dock for not taking actions to combat climate change.

I think more than laws, regulations and sermons and 'head talks for binding people' we need an approach more from our hearts and start taking actions, demonstrating 'good models' which can be multiplied several thousand fold. We must remember - 'an ounce of good action is better than tons of advice' !.

Some initiatives and action plans are listed here and discussed in full details elsewhere. These are :

The 20:20 model of Eco Agriculture i.e. higher farm production with lower input cost by eco-friendly agri-bio-inputs.

Eco village – clean village program converting agro-waste (with multi microbial mixtures) to standard and high value Bio Compost - Relife and Wonderlife.

Low Cost Green Houses and integrated farming–examples Kutch Rehabilitation program and to some extent Nepal program post-Earthquake Disasters.

Saline water Agriculture and integrated aqua–agro complexes at Arizona – Mexico, Mundra (Kutch) and Navsari South Gujarat.

Bio Shield – Demonstration projects at Kalpakkam (Tamil Nadu), Uttarkashi (Himalayas) and Pavagadh – Vishwamitri River (Central Gujarat).

Vaho Vishwamitri Abhiyan – a holistic river revival project consisting of (a) Water harvesting and water bodies (b) Bio Shield on the banks of the river (c) Solid and liquid waste treatment & recycling and (d) Eco agriculture / organic farming around the water body.

Each such project can become a model and can be repeated several thousand times through local actions. Such initiatives are key for converting 'Know-how' to the field level 'do-how' advocated strongly by Swaminathan and others. It also emphasizes the need of taking such developments to the farmers as per the last words of Norman Borlaugh when he said "Take it to the Farmers".

RECOMMENDATIONS

Water and soil are the most important areas for the livelihood and future.

1. A great need to understand the linkage :

   Development – Agriculture – Water - Climate Change.

2. Climate change threatens the livelihoods and food security.
3. Cost of Climate Change will be equivalent of loosing 5 to 20% of GDP [World Bank].
4. Agriculture sector must be placed at the center of the development agenda of the MDG (Millennium Development Goals).
5. Agriculture holds a significant potential to face the change through enhancement of sequestration and reduction of green house gases.
6. Crop improvement through traditional plant breeding and biotechnology tools to target tolerance to drought, salinity, heat, flood etc.
7. We have to also aim at the exploitation of alien genes as well as new genes for gen pyramiding, converting C_3 into C_4 plants, building multiple resistance to both biotic and abiotic stresses.
8. Restoring degraded soils and eco systems through increase in carbon pools in soils and vegetation. Application of soil amendments such as

crop residue, improved biocomposts & organic recycling. Alternative packages of agri bio inputs and water use efficiency systems.

9. Improving water use efficiency at all levels.
10. Improve soil / ecosystems, social resilience. Promote soil carbon sequestration – an effective strategy to improve soil health and raise crop yield adaptation of 4 per 1000 initiative.
11. Providing incentives to farmers for ecosystem services.
12. SOC (Soil Organic Carbon) in most soils of India is only 0.2 to 0.5 percent which is much below the threshold level of about 1.1 percent.
13. Stop burning of wheat, rice straw *etc*. and take up recycling using Multi Microbial consortia (e.g. Wonderlife 1,2,3 and Relife) – treatments to convert agro waste to bio-composts *in-situ*.
14. Holistic Plan for revival of river and water resources. Special emphasis on Eco Agri practices in the vicinity of all water resources.
15. Positive actions for building new Eco – Economy and optimism for new opportunities and a better world.

CONCLUSIONS

The problems and challenges of climate change are well recognized.

Water and Agriculture are the two most affected areas due to climate change.

Together, they two are also areas of solution.

New Eco – Economy, inventive spirit of human race will bring unpreceded opportunities to convert the big problems into big opportunities.

Already several such good examples, cases and action plans exist all over the world.

We need Actions - Actions – Actions.

REFERENCES

1. Swaminathan M.S. From Rio to Johnesberg – "Action Today – Not Promises for Tomorrow" – East West, 2002.
2. Kapil Mehan, Agri Year Book 2016 – Agri Today, New Delhi.
3. Mehta M.H., Key Note at "Sustainable Agribiotech for Africa" International Conference, Mauritius, 25 – 28 September 2008.
4. J.C. Dhingani and V.P. Chowatia -Soil Testing, Junagadh Agricultural Univ.. Gujarat Krishi Jagaran (Gujarati), November 2015.
5. Claude Bourguignon – Regenerating the soils; The other India Press, 2005.

6. Murli Gopal et al; 'Food for Thought', Current Science, October 2013.
7. Report – World Watch Institute, July, 9, 2015.
8. Soil Health Card Scheme : Editorial – Financial Express, Feb. 23, 2015.
9. Uphoff N. – SRI & CSI – Down to Earth, July 2014.
10. 4 per 1000 initiative – IFOAM, December 2015.
11. World Commission on Water Proceedings – MSSRF, Chennai, January, 9 – 12, 2003.
12. Kalam APJ and S.P. Singh – Target 3 billion, Penguin Book, 2011.
13. Mehta M.H. & Swaminathan M.S – "Rehabilitation of Earthquake affected Gujarat", GAU Publication, 2002.
14. Gupta I.C. & Gupta S.K.; Use of Saline Water in Agriculture, Science Publication, Jodhpur 2003.
15. Reddy M.P. and Iyenger E.R.R – Handbook of Plant and Crop Science; M. Publication, Arizona.
16. Personal Discussion : Hodges Carl N & Mehta M.H – Sea Water International Arizona, between 1972 to 2003.
17. Narayan Sunita, Why I should be Tolerant, Centre for Science& Environment, 2016.
18. Borlaugh Birth Centenary Dialogue – MSSRF, Chennai, March 13 – 15, 2014.
19. Lester Brown, in 'The Earth From the Air – Yann A – Bertrand, Thames & Hudson 2002.
20. Mehta M.H. : Profitably Preventing Pollution – Ankur Publications – Bombay, 1990.
21. Community Science Centre – Vadodara : "Vaho Vishwamitri Abhiyan – a holistic plan for revival of River" – CSC Report 2009.

"Our dreams are for the earth, not heaven,
because heaven is here to make, not on cloudseven"

— *Muni*

24

The Way Ahead

It is well recognized that sustainable management and sustainable livelihoods are inseparable. If neglected and if the present trends are not corrected in time, everyone's future is threated. On the other hand, there is increasing recognition of living examples and demonstrated evidences of how the eco agricultural practices can help provide a solution to some of the major problems of today and tomorrow. Also, the main stream international agencies are emphasizing that agro-ecological approaches are the way forward. What the policy makers, farmers and people would like to know is How !it has been our experience that in most countries of the developing world the answer to 'How' of eco-agricultural is being demanded!

In January 2016, a global team was invited by the office of Prime Minister of India to discuss and prepare an approach paper on Agro-ecology (or Eco Agriculture including Organic farming). The team consisted of J. Fagan, Hans Harren, M.H. Mehta and A. Thimmaiah who after intensive discussions had prepared several presentations – first given to the group of top officials of Government of India departments and institutions as organized through the office of the prime minister. This was later followed by a meeting with Prime Minister Modi. During this meeting, the clear question asked was, "You need not convince us that agro ecology and organic farming are the way forward, but the question is how? How do we expand this system to national scale ?"

BOX - 1
GLOBAL EXPERTS MEETING

AGlobal team was invited by PM (O) – New Delhi to discuss regarding Eco Agri / Organic Farming large scale initiative. The Expert group consists of

1. Prof. John Fagan (A leading Health / Food Biotech Expert – Ex-Scientist – NIH – US & now at Mah. University – Iowa USA)
2. Dr. M.H. Mehta (Chairman – The Science Ashram / Gujarat Life Sciences – Vadodara and Ex. Vice Chancellor – Gujarat Agricultural University)
3. Dr. Hans Herren – World Food Prize Laureate – 1995 and President – Millennium Institute – Washington.
4. Dr. Thimmaiah– Advisor – Bhutan Govt. on Organic Farming / Bhutan (the only/ first country to go 'Organic').

The meetings were in 2 parts : PM Office Conference Room with Secretaries & DGs etc and later with PM at PM's residence on 13.01.2016. Though the Global Expert Committee will now work out a detailed Road Map for Indian Govt. - falling out of this high level meeting are some relevant points:

1. Eco Agriculture Revolution is the global need and India is in ideal situation to take lead.
2. It will bring poverty and hunger removal, economic and environm-ental sustainability.
3. Help fulfill COP-21 commitment.
4. Successful global models are available based on eco-friendly Agri Bio Inputs, integrated farm waste management and marketing. These can be adopted and multiplied in India in a big way.
5. The 20:20 model is practical and viable for transition from conventi-onal to Eco Agriculture.

The search for a suitable answer to this question becomes the next mission . Such a question – query we have experienced in many countries. The road map is being sought by a number of well meaning heads of states of different countries.

FIVE CORE AREAS FOR ECO AGRI REVOLUTION

The five core areas high lighted in the report in the enclosed Annexure – 1. The major need for a decision is to bring about a major change in transition from the conventional to eco agricultural (agro ecological) approach and implemented in future.

The first basic action suggested is Re-allocation of agricultural resources to allow rapid ramp up of Eco Agriculture / Agro Ecology.This path does not require high allocation of fund but is evolutionary in nature. Training, introducing alternative bio inputs and a major change from the outdated conventional practice to the eco-friendly practice is the way.

It is also suggested to have a major change in the Agricultural Ministries at national and state level. For this a task force or a reorganized and empowered department of Eco Agriculture is required.

Though the above suggestions appear difficult to implement in the large Government machineries, it is achievable. Over the last two decades, we are seeing distinct changes in many state governments who allocate special budget for eco friendly inputs, policies of increasing organic farming and increasingly reaching out to the farmers with 'special eco friendly input kits'.

Also the five core areas to be emphasized to accelerate transition to 'Science based eco agriculture / agro ecology elaborated area:

1. Research
2. Education
3. Technology Transfer-lab to land
4. Infrastructure and institutional development
5. Marketing

It is necessary to work out a transformational plan in each of these areas. The major challenge of bringing such a transformation is difficult but inevitable in view of the increasing awareness at all levels, demands from farming communities and the global need.

NEW DEVELOPMENTS AND NEW HOPES

In the debates conventional verses eco-agricultural, some people often 'quote' (or 'misquote') Norman Borlough, say that if you take away chemical fertilizers and pesticides, half the world will starve! On the other hand there are enough demonstrated evidences of efficient eco-friendly alternatives. Moving towards sustainable, productive farming that will greatly help the present and future generations. A timely transition to eco- agricultural future is possible and desirable and some of the developments and demonstrated model which we have discussed in this book can be revisited and listed as follows:

1. Understanding and appreciation of eco- agricultural practices.
2. Developments in close to nature technologies, visible growth and applications of a) Bio-fertilizers – New consortia technology b) Bio-pesticides – Rise in global industries. c) Bio-compost d) Agro-waste utilizations with multi-microbial technology.
3. 20-20 model – Higher farm production with lower inputs cost in a sustainable manner.
4. Eco-friendly bio input packages for varieties of crops, urban agriculture, sports fields and home-gardens.
5. Eco-village-Clean-village Program.
6. Bio-shields for coastal, hilly and ravines.
7. Eco- agricultural rehabilitations for people affected by Earthquakes and Tsunami like disasters.
8. Water use efficiency – more crop for drop.
9. Holistic Projects for Revival of Rivers Water bodies.
10. Saline water agriculture.
11. ICT as valuable tool for tech transfer and reach the unreached.
12. Soils – Water and Eco-agriculture as important part of action plans.

The above 'some' examples only indicate new emerging ways. Scientists around the world increasingly recognize the urgent need to a step change in efforts in providing solutions to the challenges. Societies and farmers will demand more polices and action oriented and demonstrations and we will need breakthroughs in technology, experimental demonstrations and systems more effective and a better exchange and communications with general public will be a part of contract between science and society. We will need a new generation of integrated thinkers and doers. Learning from traditional knowledge of farmers and exploring and demonstrating the new Eco-friendly technology will be critical.

Working with farmers in different parts we experienced the two trends:

a) In the Green Revolution areas like Punjab, Haryana – after initial reluctance, now there is increasing acceptance of Eco-friendly inputs and tendency to reduce the chemical inputs.

b) In the untapped production potential areas like Chattisgarh, Bihar, Assam, North Gujarat, Tamil Nadu, Karnataka, etc. there is growing action in adopting eco-friendly inputs and the demonstrated results of bridging the yield gaps and optimism.

WORKING GROUP ON ECO-AGRICULTURE

Indian Council of Food and Agriculture – is an apex body to act as think tank, policy research platform, trade facilities and facilitation and monitoring centre for food and agriculture sector and also as a global forum for partnership.

Recognizing the significance of eco-agriculture including organic agriculture and bio products a working group has been formed.

The group has representatives from leading institutions, industries, government departments, farmers bodies and NGOs and leading scientists and economists. Key officers from central departments (and some state governments) like agriculture, environment, SME (Small and Medium scale Enterprises), Regulatory bodies etc. are important part of such a program to facilitate planning and implementation. It is heartening to see the awareness and enthusiasm of all to participate in making a road map for eco-agriculture future. It will also draw a great deal from the recommendations prepared by the global expert committee and other sources like UNEP Green Economy report; IAAKS TD, Organic farming institutions and farmers organizations.

ECO-AGRICULTURE REVOLUTION FOR A BETTER WORLD

We are all deeply concerned and often frustrated when we think of the undesirable and unsustainable scenario of earth. We are the first generation to know that we have become threat to our own existence as a modern civilization. We humans, Anthropos in Greek, have become such a driving force of global environmental change. With recklessness we have created of our own Geological Epoch, the Anthropocene.

Still, despite all obstacle and gloomy picture, we have reasons to be optimistic. The first reason for this is new knowledge. We know and appreciate the threat to our future and the denial of that is no more an option. Amongst all the problems of agriculture and environment, we know there are hopes. The emergence of new eco-friendly development and demonstrated success stories need to be scaled up and should reach the smallest farmers.

Eco-agriculture will take us to a world where we are a part of nature, instead of 'apart' from it.

REFERENCES

1. Rockstram John &Klin Mathias – 'The Human Quest,Langenskiold, Sweden, 2012.
2. Agro Vision, Fergussion& Co., and GAU Documents; 2002.
3. Grutzen, P.J. and Stoermer E.F., ' The Anthropocence', Global Change News letter 41:17-18, 2000.

ANNEXURE

CONCEPT NOTE

Scaling Up Organic Agriculture and Agroecology throughout India

J. Fagan, Ph.D., H.R. Herren, Ph.D., M.H. Mehta, Ph.D., and
A. Thimmaiah, Ph.D.
22 February 2016

> *"India's journey to prosperity can be a more sustainable and environmentally sensitive one than the path followed by countries that came of age in earlier eras. This is a journey of our choice, rooted in our tradition that worships nature's bounties."*

—Prime Minister Narendra Modi, Wall Street Journal, 25 September 2014

Executive Summary

This concept note rests on the scientific evidence that we presented in detail on 13 January 2016 during meetingswith Prime Minister Sri Narendra Modi and with key members of India's agricultural leadership.

The core conclusion emerging from those meetings is that, at this point in time, "business as usual is not an option." Instead, a timely transition to sustainable organic agroecological approaches is essential to protect the security and well-being of the Indian people and India's farmer majority, as well as the population of the world as a whole.

At this time, the weight of scientific evidence has swung definitively toward the conclusion that organic, agroecological, regenerative approaches to agriculture are the scientifically best documented approaches to take to address the needs of India's agricultural system and its farmers.

Organic and agroecological methods deliver increased productivity, higher quality, and more balanced nutrition, and they build the fertility of the soil and protect the ecosystem. These methods are also in step with the simple reality that, even though fossil carbon-based agriculture has been useful in the short term of the last few decades, this approach is highly inefficient (wasting 10 calories of fossil carbon for every calorie of food produced) and, because of the progressive depletion of oil and gas, this limited resource, cannot sustain humanity into the future. Instead, it is necessary to look to organic and agroecological approaches that make use of renewable on-farm resources to manage fertility, pests and disease in a sustainable manner, and that protect the environment, preserve essential ecosystem services, and feed the people of the nation reliably, securely, and abundantly.

It should be acknowledged that, there are sustained efforts by entrenched interests to counter this conclusion. However, the weight of scientific evidence is clearly in favor of organic and agroecology. In our meetings we presented the essential elements of that scientific evidence.

We were very happy to hear the following message in our meetings with Prime Minister Modi and with the agricultural leadership of India: "You need not convince us that organic agriculture and agroecology are the best way forward, but the question is *how*? How do we expand this system to national scale?"

This concept note was prepared to answer this question, and suggests the following four steps:

First, a definitive decision needs to be made that business as usual is not an option, and that a transition to organic and agroecological approaches should be decisively implemented in a timely manner.

Second, following on from that decision, two lines of action need to be taken:

1. Funding needs to be redistributedto emphasize organic agriculture and agroecology in all agriculture and food related government institutions and programs, so as to enable the rapid ramp-up of this approach.
2. An Organic and Agroecology Taskforce needs to be created within the Ministry of Agriculture. The Taskforce would be responsible for guiding the transformation of the administrative, research, education, extension, infrastructure and marketing frameworks of the Indian agricultural system. It would also guide this system to continuously evolve, operating progressively more sustainably, and reflecting to progressively greater degrees regenerative, organic and agroecological principles and practices.

Third, the Taskforce needs to take a fresh look at the Indian agricultural system, using currently available assessment and modeling tools, and on the basis of that assessment, work with stakeholders to create, and vigorously implement, an action plan for transitioning to organic and agroecology on a national scale. We provide a concrete plan for this process.

Fourth, given the timely nature of this matter, it would be worthwhile to implement certain key interim measures quickly, even during the time when the in-depth action plan, described in the previous paragraph, is still in development.

These interim measures would be undertaken in the following five core areas:

1. **Research**—New research is needed to quickly adapt international research findings in organic and agroecology such that they can be applied effectively in the Indian context. Additionally, farmers need to be engaged

directly in setting research objectives and carrying out research. This will greatly reduce research costs and will accelerate knowledge transfer "from lab to land", because the research will already be in the farmers own fields.

2. **Education**—It is needed to train large numbers of new experts, and retrain all currently active agricultural technologists in organic and agroecological principles and practices. This expertise needs to be available at a scale that will enable this knowledge and technology to be quickly transferred to every farmer in India.
3. **Extension**—Newly trained and re-trained agricultural technologists need to be systematically and widely deployed within Krishi Vigyan Kendra, the agricultural extension system, such that the knowledge and technology of organic and agroecology can be developed with and delivered quickly to every farmer in India.
4. **Infrastructure Development**—Infrastructure and institutions need to be developed and/or reorganized to support the activities of the farming community, especially this needs to be done in a manner that facilitates, encourages and speeds the transition to the organic and agroecological farming methods.
5. **Marketing**—Marketing structures and specific initiatives need to be created that enable farmers to reap fully the rewards linked to transitioning to organic and agroecological practices. This is needed in order to maximally improve the economic and social welfare of each member of India's farmer majority. This is also needed to transform India into an international agri-trade powerhouse, producing premium, high-demandproducts that fetch high prices based on organic and Vedic organic branding, and deliver abundant profits to all players in the Indian agri-supply chain. This will contribute strongly to strengthening and augmenting India's economy. Whether called education or marketing, it is also necessary for consumers to be informed more fully of the importance of good nutrition for health, and for them to understand that organic delivers food having higher nutritional value, as well as benefitting the ecosystem. This can be done through schools, public food services, etc.

We are confident that if action along these lines is undertaken in a timely manner, profound, smooth and progressive transformation can take place in India's agricultural system that will ensure a secure, abundantly prosperous future for India, her people and her farming majority, and protect the traditional cultural values of the nation.

Introduction

In meeting with Prime Minister Modi and India's agricultural leadership on 13 January 2016, we were gratified to find a shared appreciation of the need to transition India to regenerative food production systems based on organic and agroecological principles.

This shared appreciation is in tune with the growing enthusiasm for organicagriculture and agroecologyrising in India today. This enthusiasm is partly because this approach is in harmony with the cultural values of the Indian people. Pragmatically speaking it is also a reflection of the effectiveness and success of this approach, documented by extensive scientific researchand an increasing number of examples of successful organic and agroecological farming initiatives emerging in every state of India. [1,2,3]These testify to the fact that this approach consistently delivers substantial and multiple benefits: stable increases in agricultural productivity, efficient use of local on-farm resources, improved nutrition and safety,[4,5,6] and environmental regeneration.[6,7,8,9]Finally, this enthusiasm is in line with the growing recognition that India needs an agricultural model that addresses holistically the multiple challenges ahead.

During the meeting with Prime Minister Modi, we were asked the following question, "You need not convince us that organicagriculture and agroecology are the best way forward, but the question is *how*? How do we expand this system to national scale?"

In answer to this question, we suggest two basic of actions.

1. Re-allocate agricultural resources to allow rapid ramp-up of Organic Agriculture and Agroecology

The key action required is to redistribute funding to emphasize organic agriculture and agroecology in all agriculture and food related government institutions.

We congratulate the government for the very important programs that have already been implemented, most notably Paramparagat Krishi Vikas Yojana, for expanding organic agriculture nationally, and for the other very important agricultural initiatives already in progress.[10]In addition, we would encourage an even stronger approach to ensure that the nations stated agricultural goals are achieved.

Given the current circumstances of India's farming majority, business as usual is not an option. What is needed is a decisive and resolute decision to adopt the scientifically most developed agricultural system available today—advanced organic agriculture and agroecology[11]. Allocating robust funding to quickly expand these technologies to national scale will make it possible to better nourish the Indian people, expand the economy and international trade, and achieve the agricultural and sustainability goals India has agreed to under the UN Sustainable Development Goals (SDGs).

Rapid expansion of a biodiverse production is needed to self-sufficiently and sustainably nourish the nation for the medium and long term. Ramping up organic agriculture is the fastest and least expensive way to accomplish this. As we reported in our presentation on January 13th, research in many countries around the world shows that providing farmers with training in scientifically advanced, highly productive and efficient methods of organic agriculture consistently leads to substantial increases in crop yields.[12,13,14,15,16,17] In many cases, especially with small-holder farmers, whose production systems are not highly optimized, training in organic farming techniques can double crop production sustainably in just one or two years.[12,13,14,15,16,17,1819]Similar results have been reported in India. [1,2]

The power of this approach is that it does not require massive investment in new technology. However, the primary action required is to provide funding to re-train each individual member of the existing farmer majority to apply more efficient, science-based organic methods on their own farm. Farm by farm, this training strategy can lead to a massive increase in national agricultural productivity, in principle, approaching a doubling of national production. The resultant increases in productivity would be more than sufficient to address the nation's needs.

It will also be necessary to provide funding for improvements in infrastructure and farmer support, so that the increased production can be stored and distributed effectively and sold profitably.

These two investments are a fraction of the cost that would be incurred in extensive implementation of chemical-industrial agriculture. Moreover, while implementation of chemical-industrial agriculture would entail massive social, cultural, and economic disruption, wide-scale implementation of organic agriculture would strengthen and stabilize the traditional rural culture of India.

These two relatively modest investments in training and infrastructure will pay for themselves within a few years by improving national food security, improving farmer incomes and advancing both the local and national economy.

2. Create an Organic and Agroecology Taskforce within the Ministry of Agriculture

The second line of action required is to create a taskforce within the Ministry of Agriculture, responsible for guiding, over a period of 5-10 years, the transformation of the administrative, research, education, extension, infrastructure and marketing framework of the Indian agricultural system, such that this system operates progressively more sustainably, reflecting regenerative, organic and agroecological principles and practices, and continues to develop along the path of sustainability. This task force would report directly to the Prime Minister, would have broad representation, and would be given the mandate to quickly put in place the specific policies and programs needed to enable a steady, nation-wide shift to organic agriculture.

Creating leadership that has (1) this mandate, (2) the experience and knowledge of organic agriculture, (3) the necessary strategic planning, management and leadership skills and (4) access to the necessary financial resources should be sufficient to achieve the goal.

However, we wish to take the opportunity to suggest the outlines of a road map for how this goal could be achieved in the pages that follow.

National Agricultural Assessment and Action Plan

In recent years, powerful assessment and modeling technologies have been developed to support comprehensive, evidence-based planning on the national, state, and local levels. It would be very fruitful for the new organic agriculture Taskforce to undertake a national- and state scale assessment of the agricultural system.

This would enable India's leadership to take a fresh look at the national agricultural system and work with all relevant stakeholders to design a robust, comprehensive regenerative agricultural action plan for the nation that optimally incorporates the principles and practices of organic agriculture and that is in line with the nation's vision for sustainable development.

The process for undertaking the national and state assessment and for developing the action plan would include representatives of all stakeholder groups in the India's agri-food system. These would include government policy makers, government and academic researchers, farmers and consumers, as well as, environmental, human rights and agricultural NGOs, farm cooperatives, other farmer associations, public health and health-care experts, development partners and representatives of various sectors of the agri-food industry.

The process for carrying out the national and state assessment and developing the action plan would take into consideration all on going activities in the agricultural sector across the spectrum of research, education, extension, infrastructure and institutional development, and marketing, with the goal of identifying and evaluating different options for transforming India's agricultural system.

The national assessment and development of the action plan would unfold through the following four-steps:

1. Model development

Guided by existing data resources and extensive input from the wide range of stakeholders, described above, a computer model can be built,using a platform such as the Millennium Institute's Threshold 21platform[20]. The model can be customized to take into consideration all dimensions of the Indian agri-food system and the broader social, economic, and environmental system within which it exists. This could also be done at other geographical scales—states, communities, cities, and this model could also serve as a tool to inform policies relevant to the full range of the UN Sustainable Development Goals.

2. Scenario creation and analysis

Once validated according to standardized procedures, the computer model can serve as a powerful but easy-to-use tool that each of the stakeholder groups in India's agri-food system can employ to explore the potential impacts of specific policy choices for India's people, economy, and environment. In essence, the model enables stakeholders to ask the question, "If policy X is implemented, how will that influence parameters A, B, C,... over Y period of time?"

Capacity development and technical support can be provided to enable all stakeholders to use this tool to explore in depth the ramifications of the full range of policy options applicable to Indian agriculture.

3. Defining an action plan

Based on this process of modeling and scenario analysis, stakeholders will gain a clear understanding of the ramifications of specific policy alternatives.With this understanding, a process can be facilitated through which the wide range of stakeholders in the Indian agri-food system can work together to select specific policy alternatives to achieve a set of consensus goals. Based on these, a national agricultural action plan can be developed.

4. Implementation, monitoring, evaluation and course adjustment

Implementation of the action plan entails continuous cycles of monitoring, evaluation of results, and course adjustment to bring outcomes into line with objectives, and ultimately achieve the consensus goals established by all stakeholders. An overview of this process is presented in the figure, to the right.

Some examples of scenarios and policy options that could be explored as part of the modeling process

A key part of developing an action plan is testing various scenarios, as described in Step 2, above. In the paragraphs below, we identify policies and programs, in each of the areas mentioned above—research, education, extension, infrastructure and institutional development, and marketing—that could be evaluated in this way.

Some of these scenarios also represent possible modifications to India's agricultural system that could be implemented even before the strategic planning process is complete, thereby accelerating the transition to science-based organic agroecology.

1. Research

Shifting research and extension funding from the present unsustainable agriculture model to organic and agroecological practices can be a powerful force in jump-starting nation-wide implementation of regenerative agricultural strategies. The following two lines of research are essential:

a. **Adaptive research**—Globally, there is a large body of scientific research demonstrating the effectiveness of organic agriculture and methods in increasing the productivity and improving resilience of the farming system.[12,13,14,15,16,17] Applied research is now needed to adapt these findings for application in the diverse agro-ecosystems of India.

b. **Farmer-participatory research**—Engaging farmers directly in research will accomplish two things. First, it will save money by reducing research costs, and, second, it will achieve results faster, because the R&D will already be in the farmers fields. This approach will, not only, facilitate knowledge transfer "from lab to land", but will also assure that research is relevant to farmers real-world problems by creating feedback loops from "land to lab", from farmers to researchers. Participatory seed breeding programs, research on bio-control strategies for pests and diseases and on strategies for using renewable resourcesto improve soil fertility and soil water retention (and thus drought tolerance) would all be of great relevance.

2. Education

Bachelors and Masters degree programs in organic agricultureneed to be integrated into the curriculum ofevery state agricultural university.Education and training are key to rapidly expanding organic agriculture in India. When small-holder farmers are trained in advanced scientific organic farming methods, yields increase, often by two-fold or more, and productivity becomes more resilient to adverse weather and pest conditions.[1,2,12,13,14,15,16,17]

These universities can generate graduates qualified to become new extension agents and, through continuing education, can retrain existing extension agents, as well. This will create a pool of knowledgeable extension agents who have the knowledge necessary to train farmers to adopt science-based organic farming methods.

3. Extension—"Lab to Land" (Technology transfer)

The Indian Council ofAgricultural Research must be congratulated for creating Krishi Vigyan Kendra, which has units in every district of India. By training KVK personnel in the scientifically most advanced knowledge of organic agriculture, they can become the prime actors who can accelerate the transition to organic by sharing this knowledge with every Indian farmer.

This can be accomplished by both hiring new graduates of the bachelors and masters programs in organic agriculture and agroecology, and by retraining existing extension agents trained, as described in Section 2, above. Extension agents trained in this way will be capable of very effectively delivering knowledge ofthe powerful, science-based knowledge of organic and agroecologyto India's farmers.

a. Knowledgeable extension agents in every village

The KVK extension service needs to be redirected to give each farmer in India ready access to an extension agent knowledgeable in organic agriculture. The educational programs, described above in Section 2, will provide these experts, either by training new students or retraining currently employed extension agents.

Extension agents would work with individual farmers and local farmer producer organizations (FPOs) to accomplish the following:

- Advise individual farmers—how to use scientific, organic methods to handle specific pests, diseases, crop management issues and soil fertility challenges.
- Provide short hands-ontraining courses inorganic practices and on how to efficiently and economically convert to organic farming.
- Assist FPOs in developing organic research and training programs, collective production of organic compost, bio-fertilizers, etc., collective production or procurement of organic seeds, collective development and operation of facilities for storage, distribution, processing and value-addition and collective marketing.

b. Refocus DD Kisan, the Farmer TV Channel on organic and agroecology

Organic agriculture is already covered to some extent on DD Kisan. However, these need to be emphasized more strongly. Ideally there could be a 24 hours-per-day television channel whose core mission would be to support the national transition to organic agriculture and give farmers the knowledge, tools and support needed to undertake that transition. Programming elements could include:

- Key messages about organic agriculture
  - Scientifically most advanced form of agriculture,
  - Consistently delivers increased productivity, higher yields,
  - Reduces production costs,
  - Delivers safer and more nutritious products free from toxic chemicals,
  - Delivers consistently better prices,
  - Demand is growing rapidly and consistently.
- Step-by-step how-to courses, tutorials
  - How to convert the farm to organic agriculture,
  - How to produce each of India's most important crops organically,

- Information on where farmers can receive guidance in converting their farms to organic agriculture,
- Contests and Awards—largest turnips, the most delicious apples, the highest yield of rice, the most integrated and productive organic farm, the best compost, the best bio-pesticide for a given crop or pest, etc.,
- Farmers sharing their experiences—how they produce their bio-fertilizer or deal with a pest, etc.,
- Inspirational programming for the farm family.

c. Refocus the mKrishi mobile app extension service on delivering the knowledge and technology of organic

The mKrishi mobile app is successfully delivering agricultural information to farmer families.[21] This system could be extremely useful in the transition to organic agriculture by making information about this powerful, science-based system of agriculture highly accessible to every farmer. All of the kinds of information delivered to the farmer through DD Kisan, could also be delivered through mKrishi. Other mobile apps such as the iCow [22], which doubled milk production in areas of Kenya.

d. Refocus the soil health card initiative on giving advice that speeds farmers' transition to organic

The currently operating Soil Health Card[23] program is a brilliant strategy for educating farmers about soil health. However, the soil assessment that is currently part of the Soil Health Card program focuses on soil parameters relevant to chemical agriculture and the recommendations provided to farmers are limited to prescribing which chemical fertilizers and chemical amendments the farmer should apply to their fields. By just expanding the soil assessment to include measuring parameters such as soil organic matter, water infiltration rate, and microbial biodiversity of the soil, and by expanding the recommendations to include organic bio-compost, bio-fertilizers and bio-amendments, the farmer will learn a great deal about this more sustainable and highly scientific agriculturalsystem and begin restoring their soil to resilient health and fertility. Should improved soil fertility management be widely adopted, the increased carbon sequestration would go a long way toward satisfying India's SDG emission reduction commitments.

4. Infrastructure and institutional development

The government has undertaken many laudable programs to support the agricultural sector and improve the lives of the Indian farmer majority. Small adjustments to these existing programs can powerfully motivate farmers to make the transition to organic agriculture. Even quite minor adjustments can make a big difference. The following sections provide a non-exhaustive series of examples:

a. Apply organic practices to make the Pradhan Mantri Krishi Sinchai Yojana irrigation program more efficient.

The government can be congratulated for its Pradhan Mantri Krishi Sinchai Yojana program, [9] which is already bringing irrigation to more and more farmers across India. By training farmers to use organic practices to improve water use efficiency, the efficiency of irrigation will be improved. This will save water and energy, and in addition will speed the transition to organic farming, thereby increasing the resilience of the agricultural system. More efficient irrigation—per drop, more crop.

b. Give farmers, who adopt organic agricultural practices, favorable conditions for crop insurance and crop loans.

The government's crop insurance and crop loan programs are very important for every farmer. When this safety net is strengthened, farmers will gain the confidence they need to adopt new, more sustainable organic agricultural and practices.

Giving farmers, who adopt organic agricultural practices, favorable conditions for crop insurance and crop loans, would significantly stimulate the transition to these practices. This can be justified because these practices lower risk of crop failure, and because the transition to organic will pay for itself by (1) reducing agriculturally-related damage to the ecosystem and loss of ecosystem services, (2) preserving non-renewable resources, (3) improving the economic welfare and health of farm families, and (4) helping meet India's COP21 commitments and Sustainable Development Goals.

c. Improving access to storage, processing, and value addition facilities and to distribution systems

Government investment in such facilities and systems sends a strong message to farmers that the organic farming opportunity is secure and real.

With these facilities in place, products that would otherwise be wasted will reach profitable markets, greatly benefiting the financial welfare of the farmer. Likewise, value-added processing will further increase returns to the farmer. These facilities will also improve access of all citizens to a better, more diverse and nutrient-dense diet.

It is essential that farmers, including women, be engaged directly, through their farmer producer organizations,in creating and managing storage, processing, value-addition and distribution facilities. This will improve efficiency and will greatly improve farmers' economic welfare by eliminating middlemen. Enabling collaboration between the central and state governments and farmer producer organizations and cooperatives in creating and managing these facilities is also necessary to optimize use of resources, eliminate redundancy, waste and inefficiency, and minimize costs.

d. Train farmers and farmer coops to produce bio-inputs from on-farm renewables

Organic bio-fertilizers and bio-pesticides, produced from on-farm renewables and agro-waste, are essential to organic agriculture. Training and extension programs need to be provided to enable farmers to produce these inputs, themselves, on-farm. It is also needed to support farmer-producer organizations in developingfarmer-operated production of bio-inputs.The capacity to produce these inputs on-farm and at the village level reduces the cost of inputs to the cost of the farmer's labor alone. This can transform the economics of the farm.

Disposing of agro-wastes by converting them to green inputs is, not only more sustainable, but also fits in very well with Prime Minister Modi's *Swatch Bharat (Clean India)* program.

Commercial organic inputs, if verified to be of high quality[24] can also serve a useful purpose, helping jump-start the transition to organic by enabling farmers to quickly replace chemical inputs with commercial organic inputs. As farmers become more experienced, they will naturally begin to produce their own inputs at much lower cost.

Government support for the commercial production of organic inputs can come through development incentives, rebates, grants, and tax breaks, offered, for instance, through the *Make in India* program.[25]

Commercial production of organic bio-inputs, by companies such as Gujarat Life Sciences [26], can become a strong stream within India's developing industrial biotechnology sector. Through this and other streams, India can become a world leader in sustainable, industrial biotech, converting agro-waste and other renewables into valuable products, such as organic bio-inputs, for sale both domestically and internationally. Industrial agricultural biotech is already much larger and commercially more promising than genetically modified (GMO) seeds, and should be strongly favored, since this branch of biotechnology is lower risk, and produces products that are safer and subject to much less regulatory red-tape both domestically and internationally.

e. Government support to adapt advanced technologies that improve the efficiency and ease of organic and agroecological farming

The government has begun to provide support for research and development of advanced farming technologies such as:

- GPS-guided irrigation systems,
- Cloud-based field mapping of farm information to optimize resource utilization,
- Geospatial applications using satellite data for soil analysis and for mapping atmospheric patterns for forecasting and risk management.

Designing these applications to specifically support needs of organic farmers will increase efficiency and productivity, optimize use of resources, and minimize risk. Integrating high-tech hardware into organic farming systems will also reinforce the fact that these farming systems are, scientifically, at the cutting edge of agriculture.

f. Expand organic demand by requiring government institutions to buy local organic produce

Government institutions, such as hospitals, schools, universities, research institutes, administrative agencies, and military facilities are among the largest consumers of agricultural products nationally. Requiring them to use organic vegetables, fruits, dairy, and other products, to the extent that these are available, will create long-term demand forthese crops. This can also be instituted in government-run food distribution channels.

This will strongly stimulate the transition to organic agriculture and will also make organic foods accessible to a much larger portion of the population. Together with Marketing (point 5. below), it will represent the pull element for the transformation, while new R&D, education and extension initiatives represent the push side.

Making organic products standard in government facilities, will motivate private schools, universities, hospitals, hotels and restaurants to do the same.

Generating demand in this way will not only increase the economic well-being of farm families, but will also improve their health by eliminating exposure to toxic agrochemicals, which are currently responsible for a substantial portion of rural cancer and birth defects. This will also improve health of the local population because it will reduce exposure to carcinogenic, teratogenic and toxic agrochemicals and improve the nutritional value of the foods that they eat.

5. Marketing

A national program for marketing Indian organic products, both internationally and domestically, is key to the national transition to regenerative organic agriculture.

Domestically, organic represents a large, yet-untapped opportunity. The primary market for organic foods in every nation is the middle class, and India's is huge, larger than the entire population of Europe or the US. In India, the organic market is growing at 20-25% per year[7,8,9]. However, investing in dynamic marketing could greatly accelerate demand, and stimulate much more rapid growth.

Globally, a major shift is taking place—the demand for organic food is rising dramatically, as described in *Fortune*, May 21, 2015, USA's premier business magazine, in an article entitled, *The War on Big Food.* It explains that, "the top 25 U.S. food and beverage companies have lost an equivalent of $18 billion in market share since 2009 as shoppers stewart to fresh and organic alternatives."

This shift translates into a very large international market for Indian organic products, large enough to fulfill all of India's aspirations for international agricultural trade.

Internationally, as well, organic represents a large opportunity. The scope of this opportunity is such that it could fulfill all of India's aspirations for international agricultural trade.

The most profitable approach for maximizing the revenues of India's agro-industry is to focus 100% on the new, rapidly expanding, highly profitable organic opportunity. This is especially the case, given the minimal growth, low price, low quality, low profit margins of the products generated through chemical-industrial agriculture.

It would be highly counter-productive to set up a system where organic agriculture attempts to "coexist" with chemical-industrial agro-commodity production. The presence of the latter also greatly increases the risk that organic products will be contaminated with toxic agro-chemicals or GMOs.

These risks are costly to manage, and would create doubt about the purity of Indian crops, which will weaken India's "clean and green" image and reduce the credibility and market value of Indian organic products. In contrast, by becoming 100% organic, India will gain a powerful marketing advantage. This strategy is already being adopted by some Indian states—Sikkim, Tamil Nadu, Kerala, and others, and by other countries, e.g. Bhutan, Russia and Denmark. The following marketing strategies will facilitate the organic transition:

a. Develop and market organic crops indigenous to India, having unique nutritional and/or functional value

Medicinal herbs and grains, like quinoa, are examples of specialty crops with distinctive nutritional or functional characteristics. Quinoa is a very healthy, high protein, low-gluten grain indigenous to the Andes Mountains of South America. Until 15 years ago quinoa was virtually unknown outside of the Andes, but it has now become one of the most popular grains in the US.

Bhasmati rice is an example of a specialty crop that Indian farmers are already benefiting from. There are probably many other unique crops, yet to be discovered among the traditional crops and seed varieties indigenous to India. Clues leading to such cropsin India include traditional knowledge sources such as the Vedic and Ayurvedic texts, the Krishi shastras and the rich and diverse regional food cultures of India. These are all goldmines of culturally unique products with interesting functional characteristics, which, if produced to premium organic quality specifications, would be highly attractive to bothdomestic and international organic markets.

Quinoa was never promoted intensively, and therefore it took more than 15 years for it to gain a large following. Dynamic marketing, focusing on both the functional characteristics and the cultural-social value of products could significantly accelerate this process.

Government sponsorship of such product development and marketing efforts could easily pay for itself in the form of improved local economy, increased tax revenues and improved quality of rural life.

b. Increase productivity and value by integrating the Traditional Vedic agricultural knowledge[27]and practices, indigenous to India, with the cutting edge science of modern organic agriculture

The traditional, indigenous agricultural practicesof the various agro-ecological zones of the nation can be restored and integrated with the cutting edge science of modern organic agriculture to deliver productivity and quality that exceeds the capabilities of either system alone. This traditional Vedic agricultural knowledge is highly valuable intellectual property (IP) unique to India that could bevery useful in advancing organic agriculture in India and abroad in three ways:

(1) Improve crop resilience and productivity, especially in the face of climate change, as indigenous seeds often possess these properties (2) improve nutritional value and quality, and (3) confer unique health-enhancing properties to the crop, as described in traditional knowledge sources. In addition, use of

these practices will result in production of culturally authentic Vedic agricultural products, which in itself will add substantial value to the products.

Super charging advanced, science-based organic agriculture with India's traditional Vedic agricultural knowledge, India's unique IP, would result in a highly productive agricultural system that adds value and market appeal. "*Vedic organic, green-grown in India*" could become highly competitive in global organic markets, and because this IP is unique to India, the markets created would benefit Indian farmers exclusively.

c. Agritourism—Sharing the experience of the traditional Vedic culture of the Indian countryside

Agritourism programs can be developed to allow farmers and rural communities to tap into the tourism revenue stream, thereby creating a new source of income for farm families and communities, and improved local economies.

Development of agritourism could also be one mechanism for enriching and encouraging the preservation of India's rich and diverse rural cultures. Including demonstrations of India's traditional Vedic agricultural practices can greatly enrich and add authenticity to these programs by exposing visitors directly to traditional indigenous knowledge and practices.

Currently, chemical agriculture threatens India's traditional agricultural knowledge system and seed legacy. Making practical use of the body of traditional Vedic knowledge by integrating it into the new Indian agroecological paradigm will help preserve that knowledge. Additionally, implementing agritourism programs in rural India will contribute even further to preserving this knowledge because it will bring it back into the awareness of the people and create new reasons for them to use it.

Agritourism can also be a highly effective way to inform consumers about organic agriculture, thereby helping to expand the market for organic products.

d. Consumer information—bringing to the people knowledge of the nutritional and environmental benefits of organic and agroecology

Every mother and father is concerned for the health of their children, and for their own health. Giving them accurate information regarding the connection between nutrition and health, and regarding how organic and agroecological products deliver better nutrition, can be a very effective strategy for motivating more rapid uptake of organic products. This knowledge can be integrated into the health curriculum of every school, even for young children. School gardens and similar projects can also be used to engage students directly in the food production process and enable them to learn about both the nutritional and environmental benefits of organic farming. Simple messages on health and nutrition in school cafeterias, public food services, and especially on social media can focus the population's attention on choosing organic foods, both for health and nutritional reasons and for the benefit of the environment.

e. Local organic verification systems

Certification systems to verify that agricultural products are produced socially responsibly, and according to organic procedures, without toxic agricultural chemicals, are essential to add value to these products and assure consumers of the authenticity of the products they are buying.

However, the formal organic certification systems that are used in international trade are expensive to set up and administer, and are a financial burden that farmers often experience as an insurmountable barrier to gaining access to profitable international markets.

This potential hindrance can be circumvented in two ways. First, for export, the government could implement a certification system for which the farmer does not pay or pays only a minimal fee. Second, Participatory Guarantee Systems can be set up locally that engage farmers directly in the verification process and thereby economically assure authenticity of their products for local markets.These strategies will benefit both farmers and consumers by both reducing costs and increasing availability of organic foods.

CONCLUSION

The weight of scientific evidence tells us that "business as usual is not an option". A timely transition to sustainable organic agroecological approaches is essential to protect the food security and well-being of the Indian people and India's farmer majority.

During our meetings on 13 January 2016 with Prime Minister Modi and the agricultural leadership of the nation, a question along the following lines was expressed:

"You need not convince us that organic agriculture and agroecology are the way forward, but the question is *how*? How do we expand this system to national scale?"

We have proposed four steps in this concept note that, should address this question. If undertaken in a timely manner, profound, but smooth and progressive, transformation can take place in India's agricultural system that will ensure a secure, abundantly prosperous future for India, her people and her farming majority, and protect the traditional cultural values of the nation.

REFERENCES

[1]Bhalla, N. (2010). Raising Smallholder Food Crop Yields With Climate-Smart Agroecological Practices. World Bank.

[2]Eyhorn. F., Ramakrishan. M. & Mader, P. (2007). The Viability of Cotton-Based Organic Farming Systems in India. International Journal of Agricultural Sustainability, 5(1) 25-38

[3]Malhotra, S. (2010). Indian organic food market has begun growing rapidly. Business Today. http://www.businesstoday.in/magazine/cover-story/bt-anniversary-indian-organic-food-marketgrowth/story/201823.html

[4]Baranski, M., Srednicka-Tober, D., Volakakis, V., Seal, C., Sanderson, R., Stewart, G.B., Benbrook, C., Biavati, B., Markellou, E., Giotis, C., Gromadzka-Ostrowska, J., Rembia³kowska, E., Skwar³o-Sonta, K., Tahvonen, R., Janovska, D., Niggli, U., Nicot, P., Leifert, C. (2014). Higher antioxidant and lower cadmium concentrations and lower incidence of pesticide residues in organically grown crops: a systematic literature review and meta-analyses. British Journal of Nutrition, 112, 794–811. doi:10.1017/S0007114514001366

[5]Brandt, K., Leifert, C., Sanderson, R., Seal, C.J. (2011). Agroecosystem Management and Nutritional Quality of Plant Foods: The Case of Organic Fruits and Vegetables. Critical Reviews in Plant Sciences, 30:177–197. DOI:10.1080/07352689.2011.554417

[6]Reganold, J.P., Wachter, J.M., (2016). Organic agriculture in the twenty-first century.Nature Plants2, Article number: 15221. doi:10.1038/nplants.2015.221

[7]Government Report (2010). The World of Organic Agriculture in India, National Centre of Organic Farming, Department of Agriculture, Cooperation and Farmers Welfare, Ministry of Agriculture and Farmers Welfare, Government of India. http://ncof.dacnet.nic.in/OrganicFarming-AnOverview/TheWorldofOrganicAgricultureinIndia%202010.pdf

[8]SciTech. (2014). India Organic Food Market Forcast and Opportunities, 2019, SciTech Research, 2014. http://www.techsciresearch.com/report/india-organic-food-market-forecast-and-opportunities-2019/358.html#

[9]Bhosale, J. (13 March 2014). Why Organic Farming is Gaining Ground in India. Economic Times. http://articles.economictimes.indiatimes.com/2014-03-13/news/48189580_1_organic-farming-ecocert-organic-food

[10]Government Report (2014). Agricultural Development—Sustained Efforts and New Hope. http://agricoop.nic.in/imagedefault/Ebooks270/ebookv2.PDF

[11]Although chemical agriculture has, until recently, been considered “modern agriculture,” it is now recognized that organic agriculture and agroecology are at the leading edge of scientific farming. Organic agriculture and agroecology employ the same farming techniques and technologies. Both use no synthetic inputs, such as chemical fertilizers and pesticides. Instead, fertility, pests and diseases are managed using a combination of sophisticated practices and inputs that make use of the most advanced scientific knowledge relevant to agriculture, including entomology, agronomy, plant physiology, soil biology and chemistry, plant genetics, meteorology etc. to manage fertility, pests and diseases and maximize quality, nutritional value and resilience of the resulting crop. Importantly both strive for self-sufficient, on-farm or local production of all fertility and pest management inputs from renewable on-farm resources and have a strong focus on soil health.

Organic and agroecology differ, not in the farming methods used, but only in the fact that organic farms undergo an independent verification process that independently verifies their exclusive use of the practices described above. This independent verification adds value to the product, improving the financial returns of the farmer.

For brevity, we will use “organic agriculture”, “organic farming”, or “organic” inclusively to indicate both “organic agriculture” and “agroecology”. There are many other synonymous terms, such as, eco-agriculture, regenerative agriculture, biodynamic agriculture, etc., which describe agricultural practices or systems that largely overlap with organic and agroecology.

[12]UNEP. (2011). Green Economy Report—Pathways to Sustainable Development and Poverty Eradication. http://www.unep.org/greeneconomy/Portals/88/documents/ger/ger_final_dec_2011/Green%20EconomyReport_Final_Dec2011.pdf

[13]Mckay, B., Nehring, R. (2014). Sustainable Agriculture: An Assessment of Brazil's Family Farm Programmes in Scaling Up Agroecological Food Production. International Policy Centre for Inclusive Growth (IPC-G): United Nations; Brazil, No. 123

[14]Rodale Institute. (2012). Rodale Institute 30 Year Farm Systems Trial. http://rodaleinstitute.org/our-work/farming-systems-trial/farming-systems-trial-30-year-report/

[15]Pretty, J., Toulmin, C.& Williams, S. (2011). Sustainable intensification in African agriculture. International Journal of Agricultural Sustainability 9(1), 5-24

[16]SRI-Rice. (2014). The System of Crop Intensification: Agroecological Innovations for Improving Agricultural Production, Food Security, and Resilience to Climate Change, Cornell University, Ithaca, New York, and the Technical Centre for Agricultural and Rural Cooperation (CTA), Wageningen, Netherlands: SRI International Network and Resources Center (SRI-Rice).

[17]UNEP. (2009) Agriculture at a Crossroads—International Assessment of Agricultural Knowledge, Science and Technology for Development. http://www.unep.org/dewa/agassessment/reports/IAASTD/EN/Agriculture%20at%20a%20 Crossroads_Synthesis%20Report%20(English).pdf

[18]Reganold, J.P., Wachter, J.M., (2016). Organic agriculture in the twenty-first century. Nature Plants2, Article number: 15221. doi:10.1038/nplants.2015.221

[19]Research shows that in many cases, especially with small-holder farmers, whose production systems are not highly optimized, training in organic farming techniques can quickly deliver substantial and sustainable increasesin crop production.[1,2,12,13,14,15,16,17] This will be the case for much of India's farms. However, studies comparing optimized chemical and organic farming systems generally show modestly lower yields for organic. However, even under these conditions, organic farms are more profitable, and environmentally friendly, and organic foods contain little or no pesticides and are often more nutritious. Typically organic also delivers greater ecosystem services and social benefits. Moreover, organic systems consistently deliver better yields under drought conditions, due to the increased water-retention capacity of organic soils, circumstances that will be increasingly frequent as climate change unfolds. It is thought that the small gap in yields between organic and chemical-industrial systems will likely be closed as research and breeding programs focused on organic expand and progress.[18]

[20]Threshold 21 (T21) is a dynamic simulation tool designed to support comprehensive, integrated long-term national development planning. T21 supports comparative analysis of different policy options, and helps users to identify the set of policies that tend to lead towards a desired goal. http://www.millennium-institute.org/integrated_planning/tools/T21/index.html

[21]Kumar, A. (20 July 2011).TCS's mKrishi App Brings Farming Tips To Mobile Phones, TechInAsia, https://www.techinasia.com/mkrishi

[22]iCow app. (2016). http://www.icow.org or http://www.biovision.ch

[23]Government of India Communication. (2016). Soil Health Card Scheme. http://www.soilhealth.dac.gov.in

[24]It is critical that a system be set up for verifying the quality and effectiveness of biocomposts, biofertilizers and biopesticides. Otherwise, these inputs will lose their credibility, which could slow the national adoption of organic agriculture.

[25]Government of India Communication. (2014). *Make in India* website. http://www.makeinindia.com/home

[26]Gujarat Life Sciences. (2016). Gujarat Life Sciences Website. www.glsbiotech.com

[27]Vedic Organic Agriculture is an area in which Maharishi University of Management can contribute uniquely, since the institution has worked in this area under the guidance of Maharishi Mahesh Yogi, one of the leading authorities in the field, to help restore this element of the traditional knowledge of agriculture.

25

Epilogue

Bhuva is a small village in north-west Gujarat – India. My mother was born there. The loving people of Bhuva called me 'Bhano' (Sister's Son / Nephew) and it was my favourite place to go as a child. I have my fondest memories of Bhuva, but could not visit for several decades, until when I was once at the Junagadh Campus of Gujarat Agricultural University, to inaugurate the International Conference on New Developments in Agriculture. Bhuva was not too far and I decided to go there. Soon as I reached, a familiar face, now aged with time, recognized me from a distance and with tears rolling from his old eyes (and mine) he hugged me, calling out "MaroBhano, MaroBhano" (My sister's son, my sister's son)! Now in his eighties, it was overwhelming to meet Nanji Ata, who I always remembered being wise, educated and cheerful. He was a relatively progressive farmer with about 6 to 7 acres of land, a few cows and a buffalo. Though water was scarce and saline, he managed to make a good living and was considered a well to do man.

'Khatlas' (wooden coaches woven with jute ropes) were laid out for me and the accompanying scientists from the university. Amidst the talks of new developments and technology interventions in modern agriculture, he said something that brought out the harsh realities, commonly faced in this region of western Gujarat. "Both my sons have migrated to Surat and are labourers in a diamond polishing unit" he said softly, "It is not a healthy life. The farm is still here, but they do not see enough income out of it. Inputs are expensive, water is scarce & salty and farm productivity is becoming worthless. So they decided to leave the farms to me and went off". As we sat in a brief moment of silence, he turned to me and said, "Conferences are very good, but can you people do something to get our sons back to our farms ?"

I also recall another very different incidence in Africa. The Tororo region of Uganda, is lush green and known for its fertility – quite contrary to Bhuva. We were invited by the Uganda government to explore setting up of a Phosphatic Fertilizer plant, given the large deposits of unutilized rock-phosphate there. John, a young man, probably in his thirties, came out from a small hut and met us during one of our visits there. John came up to me, clutched my sleeve and said,

Sir, please do something here. We are starving. Please set up a factory here and give us jobs."

Both these instances are typical and from different parts of the world, but had one thing in common – farmers have problems and they want a better future. Helping them to reduce their input costs, improving productivity, getting them better markets and providing long-term sustainable solutions are the needs of the day.

The aim to double the income of farmers announced recently by the Government is laudable. It will of course need the backing of many areas like markets, crop insurance, subsidies, MSP etc. But it will be useless to talk about doubling the income or making farmers 'powerful' without healthy soils, water security, ecofriendly bio inputs, farm waste management and integrated farms. It is also necessary to ask – what does the farmer really need ? Apart from better income he wants his generations a sustainable farming future. For one thing is sure, all of us care for the future generations.

There are debates about chemical vs organic farming. Organic agriculture is described by many as the best insurance policy India can have. In fact, India holds a unique position among 172 countries practicing organic agriculture. It has 6,50,000 organic producers, 670 exporters and more than 7 lac hectares under organic cultivation. This is nearly 0.4% of total agricultural land. The Government has declared intention to provide Rs.1 billion for organic market development and Rs. 3 billion for participatory guarantee scheme and other incentives.

However, there are cautions to be taken and many complex issues to be addressed before thinking of sudden transformations. In a recent report published in Down to Earth (April 2017), it has been started that despite earning the '100 per cent organic' tag, Sikkim's transition to organic farming is yet to become a true success ! Many cases and interviews of the farmers revealed the sudden drop in productivity (in some reported cases Ginger production went down to a third of the amount they used to grow when chemical use was permitted for farming !) The availability and use of alternative eco-friendly bio inputs have been negligible. Elsewhere also, some stories from the farmers suddenly switching over to organic and failing and then reverting back to conventional again, send negative signals and does lots of harm.

With this back ground, the 20 – 20 model is the evolutionary and more appropriate way for transformation. People think that despite all the problems, chemical fertilizers and pesticides cannot be wished away overnight. On the other hand the emergence of broad based eco agriculture and new generation eco friendly agri bio inputs are also welcome developments whose time has come.

New Opportunities

Along with the challenges, there are new opportunities for the younger generations on the horizon. The growing demand of new agri bio inputs and opportunity to start many small and medium scale R & D based enterprises, the demand for experts in water management and WUE, ICT, saline water agriculture, Integrated farming with livestock and aqua–agro units, waste to wealth technologies, solar appliances will all open new opportunities like never before. New need for Urban agriculture, eco friendly home gardens and sports grounds, recyclation of waste water and bio composts will be demanding innovative entrepreneurs. Another area of expertise likely to expand rapidly as productive farmland becomes scare is that of agronomists who specialize in multiple cropping, intercropping and their eco agriculture practices. Changes in the world food economy will be substantial and will need new expertise in nutrition, food storage and transportation testing and certification and eco agrieconomics !

Finally

Eco Agriculture Revolution is coming and we need to understand the following:

First to know that it will not be a sudden change and we will need patience.

The change will be evolutionary in nature.

It should be inclusive – touching the smallest farmers.

It will be incentive oriented or showing visible and incremental benefits – reduced input costs, better market rates, sustainability.

It will be based on integrated farming, standardization of practices and products and conversion and scientific use of farm wastes.

It will increasingly create new knowledge and applications of innovations.

It will open new opportunities for expertise in areas like agri bio inputs, WUE, ICT and m-Krishi, Saline Water Agriculture and agua-agro complexes, Agri Rehabilitations, Vertical Farming, Bio Shields and River Revival etc.

The new eco agriculture or ever green revolution may not be very dramatic or fast, but already there are bright examples and favorable situations and opportunities which are visible. In this era of global change, with the immense knowledge and wisdom, we will see mind shift and planetary eco revolution.